INTEGRATED CIRCUITS
MAKING THE MIRACLE CHIP

BILL PLETSCH

Cartoon art: Josie Reid

Technical illustration: Peter Waters

Cover design: Michael Cacy

Copyright 1978 by Bill Pletsch.

Revised edition copyright 1985 by William T. Pletsch

Library of Congress Catalog Card No. 84-090578

Published by Pletsch & Associates
Austin, Texas U.S.A.

Printed in Portland, Oregon U.S.A. by
The Irwin Hodson Company, Portland, Oregon, U.S.A.

All rights reserved. No part of this book may be reproduced in any form without the permission in writing from the author, except by a reviewer who may quote brief passages in review.

ISBN 0-917927-00-1

Distributed by: PLETSCH & ASSOCIATES
                    P.O. Box 06
                    Bastrop, Texas 78602-0006

About the Author . . . .

A liberal arts graduate of Arizona State University, Bill Pletsch entered the IC domain at Fairchild Semiconductor in the late 60's. After several years in the silicon materials section at Fairchild he moved to dynamically growing National Semiconductor in Santa Clara, California. He supervised a linear operation and later set up fab operations for ECL and Schottky device lines while at National.

Moving to MOS technology at Intersil, he was wafer fabrication manager for Low Power C-MOS and MOS/LSI products divisions.

Mr. Pletsch has been a strong supporter of employee education — particularly in the IC field where educational backgrounds tend to be vastly different. It was this theme that prompted the writing of this book.

TABLE OF CONTENTS

INTRODUCTION
    Purpose, Integrated Circuit Defined 5

BEFORE WE START
    Safety, Cleanliness, Waste Products, Record Keeping... 7

PRODUCT GROUPS & TYPES 11

STARTING MATERIAL
    Semiconductors, Silicon, Crystal Growing, Slicing, Polishing 15

WAFER FABRICATION
    Oxidation, Diffusion, Deposition, Masking, Protective Circuit Coating, Electrical Test 23

ASSEMBLY & FINAL TEST 61

DEPOSITION ... REVISITED....................... 65

GLOSSARY OF TERMS 70

Why does "electronics" sound so much scarier than other jobs? Probably because we are all afraid of the unknown. And yet electronics are an enormous part of our everyday lives. From ball park scoreboards, TV's and radios, watches and traffic lights, to the computer systems that put men in space, these almost miracle-working parts are made by people just like you.

When people say "I work in electronics" what do they mean? Some people may work making batteries for automobiles, or calculators, or perhaps making microwave ovens, assembling computers, or a thousand other industrial or consumer products. In this book we are going to concern ourselves with one specific area — the manufacturing of integrated circuits.

So what is an integrated circuit?

In 1947 some scientists at Ma Bell discovered that the large vacuum tubes in older radios could be replaced by very small pieces of material called semiconductor devices. Thus the transistor age was born and the world, indeed the universe, would never be the same.

At first a single transistor was built by itself on a single "chip". Later, it was discovered that more than one transistor could be made on a chip. Soon, other components, like resistors and capacitors, were added or "integrated" with the transistors. The integrating of various components on a single chip is called an integrated circuit or "IC".

Today thousands of transistors and other components are being made on a chip no bigger than this □!

This book is designed to acquaint you with some of the concepts and techniques presently being used in making IC's. The thing that is special about this book is that you don't have to be a Ph.D. or an electronics engineer to understand it. It is written for **you** in everyday language.

BEFORE WE START...

Before we start our journey through the world of integrated circuit manufacturing, let's take a little time to talk about some of the basic items that touch upon every phase of this business:
1. Safety
2. Cleanliness
3. Waste Products
4. Record Keeping

SAFETY

Above all else your safety and that of your fellow employees is absolutely essential to the success of your company. The methods or "recipes" of operation set by the management and engineering staffs in your work area have been examined at length and have included protective measures to give both a successful product and a safe method for making that product. However, because of the types of chemicals and equipment used in making integrated circuits, it is necessary that each employee help in preventing accidents. Some rules which will help reduce accidents are:
1. Be alert and safety-conscious at all times
2. Report all potential safety hazards to your supervisor
3. Report any and all safety infractions to your supervisor
4. Report all injuries or **possible** injuries at once
5. Wipe up all chemical spills immediately
6. Follow the operating specifications that have been provided for you.

Only through conscious efforts by the company and by **you** can accidents be reduced or eliminated.

OF COURSE WE FOLLOW STRICT SAFETY CODES......

CLEANLINESS

In virtually every step of IC manufacturing a clean environment is essential. Some people will work in areas where clothing much like that of a surgeon is worn. Highly purified gases and chemicals will be used. Water, air, and tools are all super clean. The reason is simple — **the "a-number-one" reason for product failure is contamination.**

Contamination can be introduced in many forms:
 1. Simple everyday dirt or dust
 2. Hair
 3. Oil from skin (fingerprints, etc.)
 4. Using the wrong chemicals, tools, or equipment.

And in some less obvious ways by:
1. Not cleaning tools or equipment at specified times
2. Using incorrect amounts of the right chemical
3. Not changing filters at specified times
4. By not following the "recipe" (operating specification) exactly to the letter.

By controlling contamination — the number 1 enemy — we are more than halfway to making **successful** products.

SOMEHOW, I DON'T THINK JUST A HAIRNET WILL DO IT!

WASTE PRODUCTS

Another extension of cleanliness deals with the waste products of the industry and environmental protection. Because highly toxic gases and chemicals are used, safe and clean methods of disposing of the waste products must be employed. Each company, usually working with governmental agencies, has spent millions of dollars in dealing with this problem. Chemical scrubbers, filters, re-cycling equipment, etc., are now in use on a broad scale helping to insure that our industry is a leader in providing for a clean and safe environment.

RECORD KEEPING

At each step of the process, records are kept to help identify what has happened during a particular portion of the manufacturing process. Operators may be asked to record (on routing cards, tickets, run-cards, etc.) such things as:
1. Thickness
2. Resistivity
3. Cycle Time
4. Weight
5. Amount "in" and "out".

This recorded data is like a life story of the process. It is of great importance to the engineering staff in determining which processing operation can or should be modified, as well as for finding clues as to why a process operation produced very good or very poor results. It is necessary to be **complete** and **accurate** each time you are required to record information.

And in some less obvious ways by:
1. Not cleaning tools or equipment at specified times
2. Using incorrect amounts of the right chemical
3. Not changing filters at specified times
4. By not following the "recipe" (operating specification) exactly to the letter.

By controlling contamination — the number 1 enemy — we are more than halfway to making **successful** products.

SOMEHOW, I DON'T THINK JUST A HAIRNET WILL DO IT!

WASTE PRODUCTS

Another extension of cleanliness deals with the waste products of the industry and environmental protection. Because highly toxic gases and chemicals are used, safe and clean methods of disposing of the waste products must be employed. Each company, usually working with governmental agencies, has spent millions of dollars in dealing with this problem. Chemical scrubbers, filters, re-cycling equipment, etc., are now in use on a broad scale helping to insure that our industry is a leader in providing for a clean and safe environment.

RECORD KEEPING

At each step of the process, records are kept to help identify what has happened during a particular portion of the manufacturing process. Operators may be asked to record (on routing cards, tickets, run-cards, etc.) such things as:
1. Thickness
2. Resistivity
3. Cycle Time
4. Weight
5. Amount "in" and "out".

This recorded data is like a life story of the process. It is of great importance to the engineering staff in determining which processing operation can or should be modified, as well as for finding clues as to why a process operation produced very good or very poor results. It is necessary to be **complete** and **accurate** each time you are required to record information.

Product Groups & Types

IC PRODUCT GROUPS

The processes used to make IC products in one company may vary from the processes of another company but there are usually some strong similarities.

By design, there are only a few major groups or divisions in manufacturing. Some are:
1. MOS
2. Bipolar
3. Analog

Usually, these divisions have sub-divisions:

```
        MOS                    Bipolar
   ┌─────┼─────┐          ┌──────┼──────┐
 C-MOS  H-MOS  N-MOS     ECL  Schottky  TTL
```

Each sub-division (for example, C-MOS or TTL) has its own method or process for making its products. Some processes in C-MOS may be very similiar or the same as a process in TTL, but basically the processes are very different. If you were to apply the C-MOS process in making a TTL product, it would fail.

IC PRODUCT TYPES

Within each sub-division, many products will be produced. A TTL group will not make just one product. In fact, they will produce perhaps hundreds of different products.

Throughout your experience in manufacturing you will be confronted with an unending barrage of number and letter codes. These numbers and letters identify a particular product that is being made.

In the auto industry there are companies such as General Motors and Ford. Within each company there are sub-groups or divisions:

GM	FORD
Chevrolet	Lincoln
Oldsmobile	Ford
Pontiac	Mercury

Still further breakouts occur:

GM		FORD	
Oldsmobile	Chevrolet	Ford	Mercury
Toronado	Camaro	LTD	Cougar
Cutlass	Monte Carlo	Mustang	Marquis
Regency	Corvette	T'bird	Capri

The IC manufacturing companies have breakouts or divisions in much the same way. The products however instead of having names such as "Capri", Mustang", or "Camaro", have number/letter codes.

National Semiconductor		Texas Instruments	
MOS	Linear	N-MOS	TTL
MM5210	741D	TMS9900	S-481

Another difference between the auto industry and the IC industry is that Ford does not make Camaros nor does GM make Mustangs. In our industry there may be as many as fifteen companies making the same product.

The reason that number and letter codes are used is because each company has far too many products. Ford may have twenty or so different models with fancy names but National Semiconductor would be hard pressed to come up with different flashy names for the thousands of products it manufactures.

In IC manufacturing, each letter/number code represents a different product. Each product will have its own special set of operations which must be performed to make it a successfully functioning "device". And, as we progress through the book, we will see how these operations affect each other.

THIS ONE IS SO CUTE, I'M GOING TO CALL IT OTIS... OH.... LOOK AT THAT ONE.....

STARTING MATERIAL

Most steps in building an integrated circuit are done on a silicon wafer. Silicon, an element like gold, zinc, or uranium, is used because of its very unique properties. First, it is very plentiful since it is basically extracted from sand — like you would find at the beach or on a golf course! Also, silicon is a semiconductor — which means it has the ability to conduct electrical current, but can be regulated in the amount of its conductivity.

Wood is a non-conductor, it will not conduct electrical current. Copper is a very good conductor of current.

The ability of a material to conduct can be illustrated by the amount of heat which gets transferred through the material. For example, if you put a wooden spoon in boiling water for five minutes the handle remains cool because heat is not conducted through the wood. If however, you put a copper spoon into the same boiling water you would find that copper does conduct heat quite well; in fact you would burn your fingers!

GUESS WHICH HAND HOLDS THE WOODEN SPOON?

Silicon, a semiconductor, falls between wood and copper in its ability to conduct.

There are other semiconducting materials besides silicon which can and are being used. Germanium is one such example. However, silicon is the most widely used material today.

The method of transforming the silicon in sand form into a wafer — the vehicle used in the manufacturing of circuits — is very intricate. The sand is highly purified and refined to remove all other matter except the silicon (if beach sand were being used shells, dirt, and beer cans would have to be removed). The end result of this refining is approximately 99.9999% pure silicon. The silicon is then heated until it is vaporized and the vapor is condensed or solidified in the form of long cylinder-shaped rods:

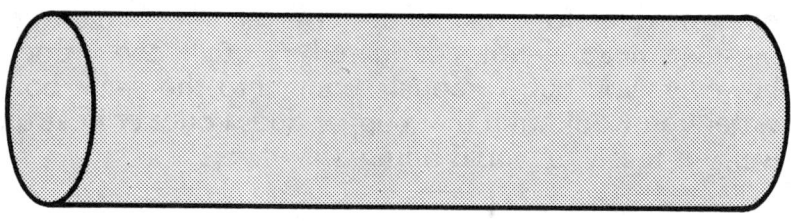

Poly-Crystalline Silicon Rod

This rod is called poly-silicon, "poly" meaning that its atomic structure is in random or jumbled order, much like a jigsaw puzzle is when you first open the box. Like a jigsaw puzzle, it will need to be arranged in an orderly fashion. This is done by cutting the poly silicon rod into smaller pieces (sometimes called a charge) and sending them to the "crystal growing" area. Here the poly is placed into a "crystal grower" and heat (approximately 1450 to 1500°C) is applied so that the silicon becomes a liquid. After the silicon is completely melted and a small amount of dopant — chemical additives that change the amount of conductivity — has been added, the actual growing process starts. A small piece of solidified silicon, called a "seed", is dipped into the top surface of the molten silicon. Because the seed is cold (solid), the very top part of the molten silicon will attach itself to it. By very delicate temperature control of the molten silicon, the amount and shape of the attaching part is regulated.

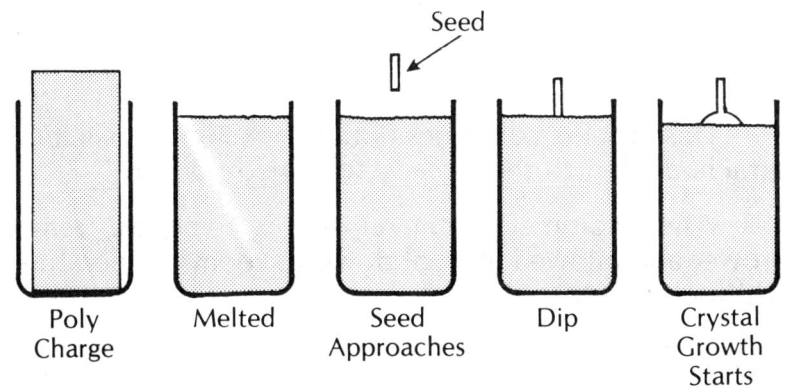

Poly Charge   Melted   Seed Approaches   Dip   Crystal Growth Starts

Starting The Growth

The seed differs from the original chunk of molten silicon in that it is **not** in random or jumbled structure. It is "mono" or single in structure as opposed to the poly charge. If the seed is single in structure, then the melted silicon, as it attaches to the seed, becomes single in structure also. This will be of great importance during the building of the circuit in the wafer fabrication area.

The growing process continues when the molten silicon attaches to the seed and solidifies. Then, more molten silicon attaches itself to the already solidified section as the "crystal" is being withdrawn slowly from the melt.

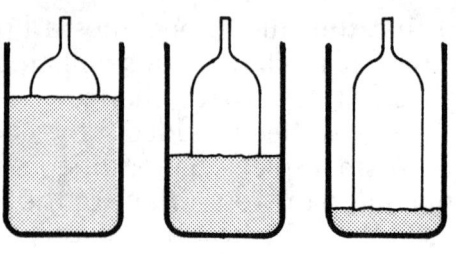

Crystal Growth

This process continues until all the molten silicon has attached itself to the already solidified portion.

When completed, the crystal is removed from the grower and allowed to cool down to room temperature. It will look like this:

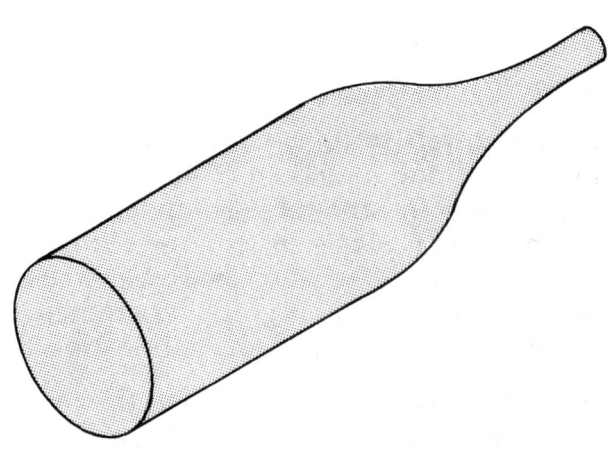

Crystal After Growing Cycle

The crystal is then "cropped" off at the seed-end and at the tail-end, and is checked for the electrical characteristics that the small amount of dopant material was to produce. It is then ground to the proper diameter. The diameter of the crystal can be made from very thin (1/2 inch) up to 6, 8, or even 10 inches in diameter. This diameter is basically controlled by temperature and how quickly or slowly the crystal is pulled or grown. In wafer fabrication, a product group may use a 4" wafer. In this case, the crystal will be grown slightly larger than 4" in diameter and then ground down to the exact size. This grinding takes care of any irregular bulges that might take place during the growth cycle and produces an "ingot" or crystal very uniform in diameter.

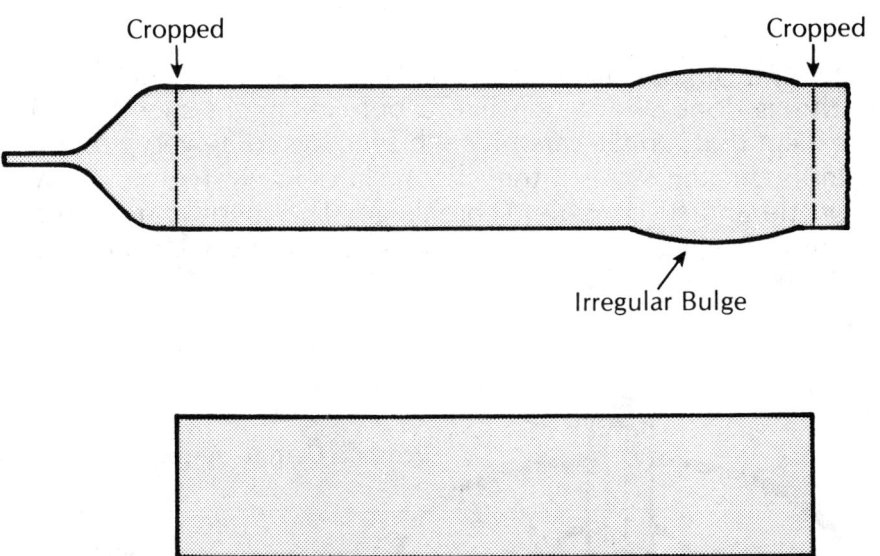

Crystal After Cropping & Grinding

Crystal Processing

Next, a "flat" is ground lengthwise on one edge of the ingot. This flat will be used as a reference point at many of the later steps of the process.

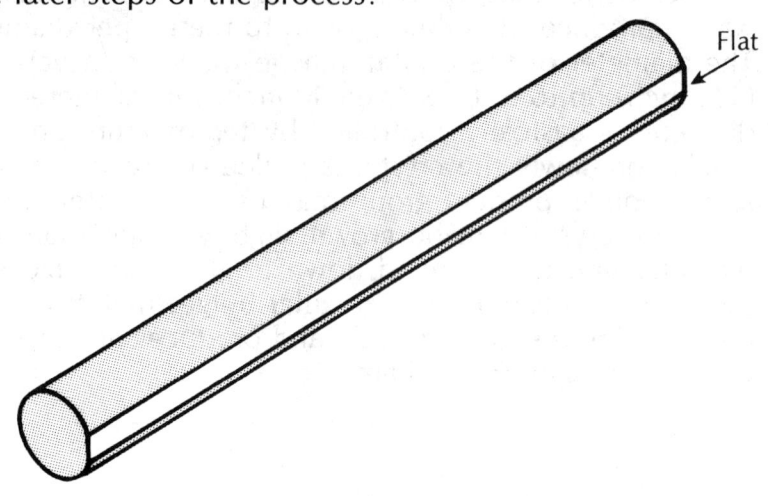

Crystal With Flat

The ingot is now ready for "slicing" into "wafer" form. This is done very much like a butcher might slice salami except that, unlike salami, each slice has to be very precise in its thickness. Also, the silicon slices (or wafers) are very brittle and susceptible to breakage. The ingot is mounted onto a sawing machine which will slice it into wafers with a diamond-tipped blade:

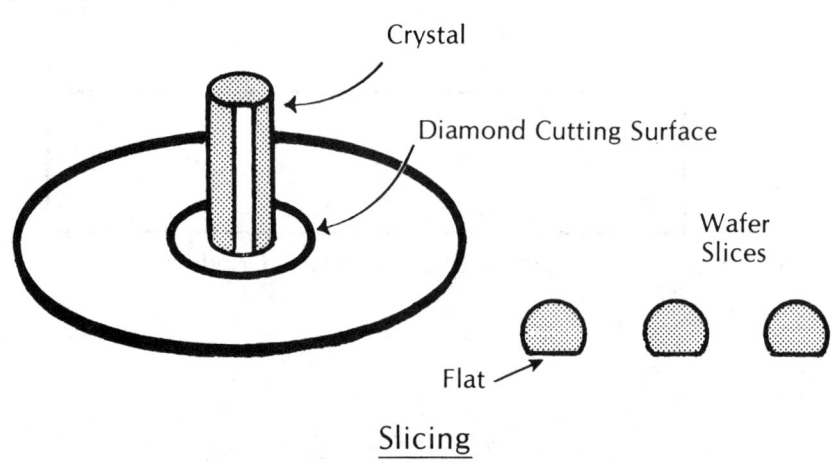

Slicing

After slicing, the wafers are collected and sent to the polishing area. Through a series of machines the wafers are lapped — ground smooth to eliminate any taper or bow that might have been produced at slicing — and are then mechanically and chemically polished so that each wafer will have a mirror-like luster. The final polish that produces the mirror like quality will be done on only one surface of the wafer (it usually does not matter which side), and it must be free from scratches and contamination of any kind.

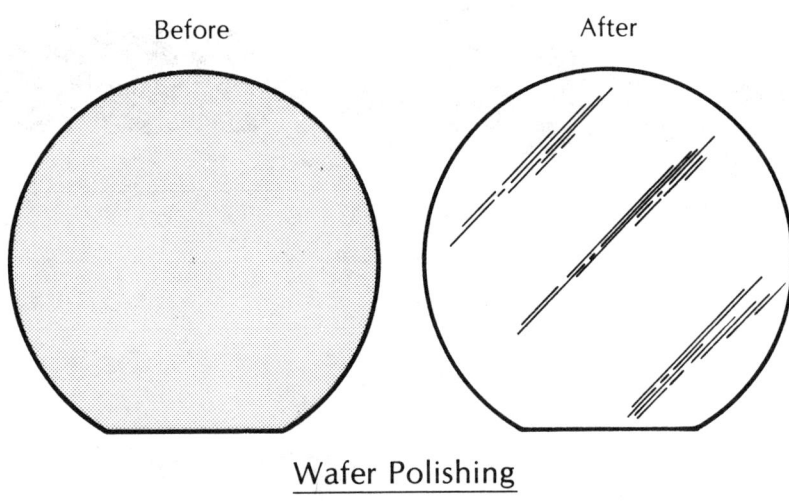

Wafer Polishing

The side on which the mirror-like finish is located will be called the front side. Most all future processing steps will be directed to this side of the wafer.

The wafers are now ready to be sent to the wafer

fabrication area where they will be transformed into working electrical circuits, the subject of our next section.

WAFER FABRICATION

At the very core of IC manufacturing is the wafer fabrication area — or "wafer fab". Because of the number of processing steps involved, more time and labor is invested here than any other area of circuit manufacturing.

Wafer fabrication involves a series of operations called "deposition", "diffusion", "oxidation", "masking", and a few others.

Some examples of work performed in these operations are:

OXIDATION	DIFFUSION	MASKING
Initial Oxidation	N + Diffusion	1st Mask
Boron Re-oxidation	Isolation Diffusion	2nd Mask
3rd Oxidation	P + Diffusion	3rd Mask
Gate Oxidation	4th Diffusion	Metal Mask

When processing wafers in fab, these processing steps are combined into a certain order or work flow that has been established by the engineering section of the department. In order to make a successful product the operations **must** be done in the order prescribed by engineering.

The "run card" or "wafer route slip" lays out the order in which these steps are to be performed. A partial run card may look like this:

Product					Run No.	
OPERATION	No. In	No. Out	Oper.	Out Date	Comments
1st OXIDATION					
1st MASK					
1st DEPOSITION					
1st DIFFUSION					
2nd MASK					
2nd DEPOSITION					
2nd DIFFUSION					
3rd MASK					
ION IMPLANT					
IMPLANT DIFF.					
3rd OXIDATION					

A run card helps identify what is happening with a "run" of wafers (wafers which will be grouped together through each step of the manufacturing process) as it is moving from one operation to the next. This particular card shows us:

1. The product type: A number or letter code is often used to describe the product and a master code can describe which exact "recipe" to use.

2. The run number: Each run of wafers has its own number. This differs from the product type in that several runs may be ICM 7038A's, but each 7038 run will have its own number starting from 1, 2, 3, and so on.

3. The operation: The steps listed below this title are in the order in which the process will be performed.

4. Number in: The number of wafers in the run when it arrived at a particular operation.

5. Number out: The number of wafers that successfully completed the operation.

6. Oper.: This identifies the operator who performed the work on the wafers at an operation.

7. Date: The date an operation was completed.

8. Comments: This can describe a part of the recipe that is used, or can be used by the operator to note irregularities that may have occurred while being processed.

A partially completed run card might look like this:

OPERATION	No. In.	No. Out	Oper.	Out Date	Comments
Product: 7215A					Run No. 343
1st OXIDATION	50	50	4175	6-1	9200 $\text{Å}^°$
1st MASK	50	49	3151	6-2	
1st DEPOSITION	49	48	6150	6-4	V/I 6.5 6.8 / NO READING

The actual processing areas also vary greatly. Sometimes an operating fab area is set up in a long line exactly according to the run card. That is, 1st oxidation will be directly adjacent to 1st mask, with 1st deposition next to that, and so on through the process.

More commonly, separate rooms will be set up grouping similiar operations together. For example, all masking steps would be performed in one room and all furnace operations would be done in a separate room. This method sounds like it is more complicated, with runs of wafers moving back and forth from room to room, however, the same pieces of expensive equipment can be used to process the wafers at many different operations. And the same operator can perform several operations at one time.

It's now time to move on and look at some of the fab operations in depth.

OXIDATION

Have you ever seen an automobile which has been sitting outside in the sun for a year or so? What has happened to the paint? It probably has become clouded and dull in appearance. Water, heat and the atmosphere have combined with the paint and formed an "oxide layer". This is basically what we want to do in oxidizing wafers. As you may have guessed there are a few added problems. First, we don't want it to take a year! It is also crucial that our oxide be very clean — and not contain any air pollutants!

As you will see throughout the book, oxide will be used as a stepping stone to most other processing steps.

Oxide formation on the car was done by rain, heat from the sun, and the atmosphere. In a similiar manner we will oxidize our wafers, but under more controlled conditions. We will substitute oxygen for the air and rain. We will use a furnace instead of the sun. Lastly, we will use the silicon wafer to take the place of the automobile.

By using purified oxygen or clean water vapor we can help guarantee that our atmosphere will be clean. With the high temperature furnace replacing the distant sun we can oxidize much faster and with more control.

In "growing" the oxide there are several things to consider. How thick do we want it to be? What else may happen to the wafer while the oxide is being grown? How clean must it be? All these questions have been considered by the people responsible for "process engineering" in your area and it is of the **utmost** importance that the recipe or **process specification** be followed to the letter. The changing or omission of any ingredient could be as disasterous to the wafers as it might be if you added a dozen eggs to a cake recipe which only called for two.

Let's go through the steps of oxidizing wafers and the importance of following the recipe should become more apparent.

Assuming that any cleaning that is required has been completed, the wafers will be loaded onto an oxidation "boat" or other apparatus which your area may use. A full boat of wafers may be from one run to perhaps twenty runs of wafers; generally "batch" processing is more economical. The fully loaded boat is now ready to go into the furnace.

The furnace is much like an oven in your home. But, there are several important differences. A home oven generally operates at up to 500°F (Fahrenheit), whereas the furnace operates at a much greater temperature; usually from 500°C (Centigrade) to 1300°C. The difference between F and C is very considerable:

Fahrenheit vs Centigrade

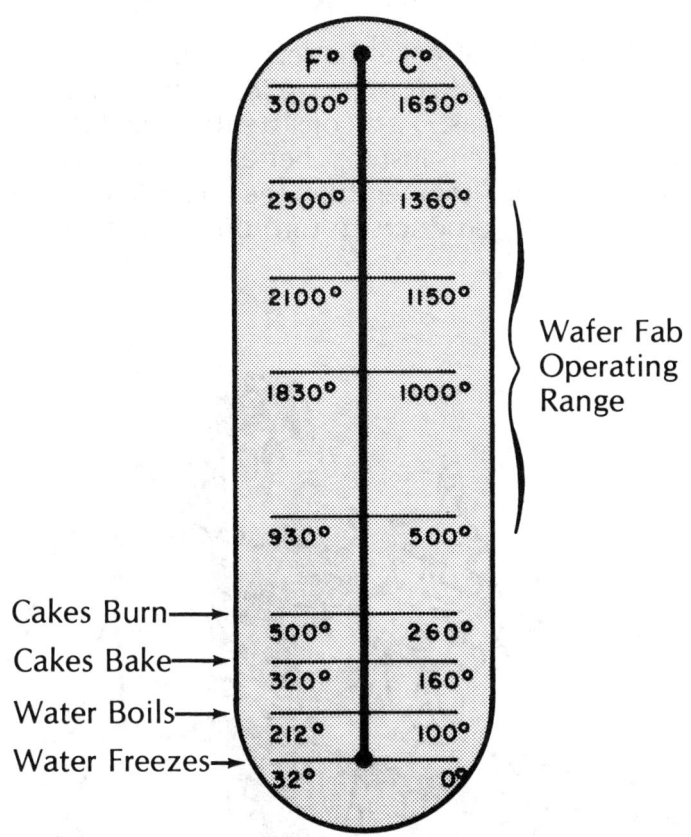

Another difference is in the accuracy at which an oven temperature is set compared to the furnace. If a recipe for baking a cake calls for a setting of 325°F, and the actual temperature is only 318°F, it won't make a great deal of difference. If, however, a recipe for a furnace operation calls for 1000°C and the actual temperature is only 993°C, it could mean the difference between success and failure.

If in baking your cake the left side of your oven is at 325°F but the right side is only 125°F, your cake is going to turn out strange at best. As a consistently even temperature throughout the oven insures better results, so must a consistently even temperature be maintained in the furnace for the IC products to turn out successfully.

While your oven at home is no doubt spotlessly clean (????), it is mandatory that the furnace environment be kept absolutely clean — remember: **contamination** is the biggest enemy!

The amount of temperature variance that is allowed is largely dependent upon the particular process on which you are working. Your engineering staff has set an operating range of temperatures which will guarantee results. This range is often set with a tolerance factor built in. So instead of only one exact temperature (such as 1200°C) being allowed, a small variance is acceptable, such as:

$$1200°C \pm 5°C \text{ (which means the acceptable range is from 1205°C down to 1195°C)}$$

The third ingredient in the oxidation recipe is "oxygen". There are several ways in which oxygen can be put into the furnace. This can be accomplished by bubbling purified oxygen through purified water (called deionized or D.I. water), or by letting plain purified oxygen flow into the furnace, or by hydrogen-burn systems. These methods vary dramatically from area to area, and from company to company. Most important, however, is what happens when we put the wafers, the oxygen, and the furnace together.

We have briefly discussed the ingredients — wafers, heat, and oxygen. Now let's look a little more closely at the other basic aspects of oxidation.

By precisely combining the ingredients we get a result called silicon dioxide or "oxide" as it is generally called. This oxide is formed by the oxygen and part of the silicon (remember, it's a silicon wafer) combining together. Some of the silicon is actually used up in the process. The major portion of the oxygen that is introduced into the furnace actually goes on through the furnace and is vented away:

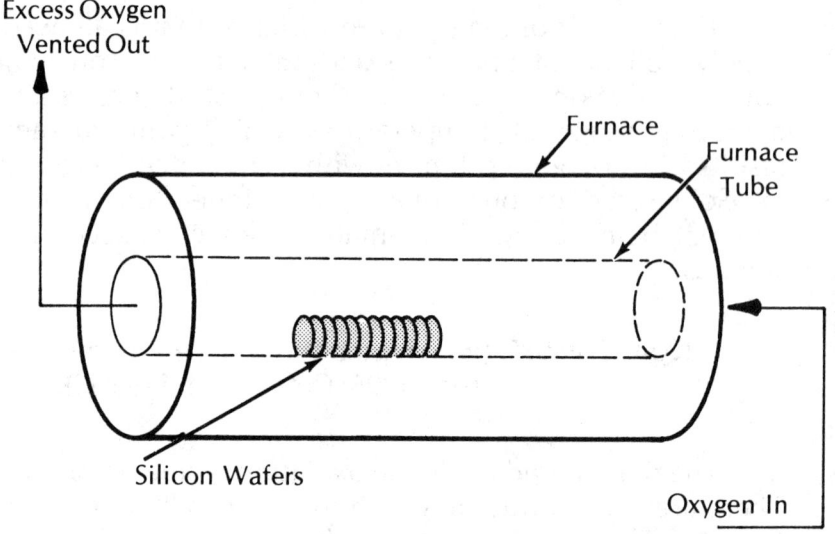

Wafer Oxidation

If either the oxygen stops or is reduced or the temperature stops or is reduced, the amount of oxide being formed will stop or be reduced.

A wafer would look like this:

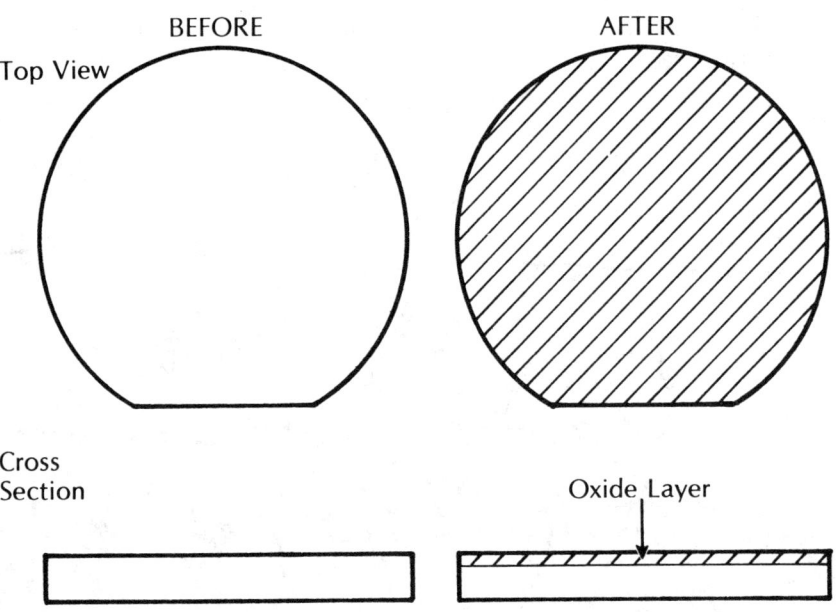

Oxidized Wafer

The oxide serves several purposes in wafer fab and in the final circuit itself. Sometimes oxide is used only as a tool for the masking operation. It becomes a vehicle for isolating certain sections of the wafer (through masking) so that later diffusions and oxidations can be put in selected spots of the wafer. In other processes it is used as a part of the electrical circuit itself, or it can be used for both of these purposes.

The order in which oxidations are done is of the utmost importance to the overall success of the circuit.

Like a surgeon performing an operation, a chef preparing a meal, or you repairing your car, it is important that not only the proper ingredients or parts are used but that they are used in the correct sequence. The key is to remember that most processing steps do more than one thing to the product and that the recipe has been painstakingly worked out to coordinate with all other processing steps.

YOU REALLY CLOSED HIM UP NICELY, MISS JONES.... HAVE YOU SEEN MY GLASSES?

The major goal of oxidation is to achieve a clean, uniform thickness of oxide on each wafer that goes into the furnace. The oxide should be the same from one side of the wafer to the other, and should be this way on all wafers on the boat. An example:

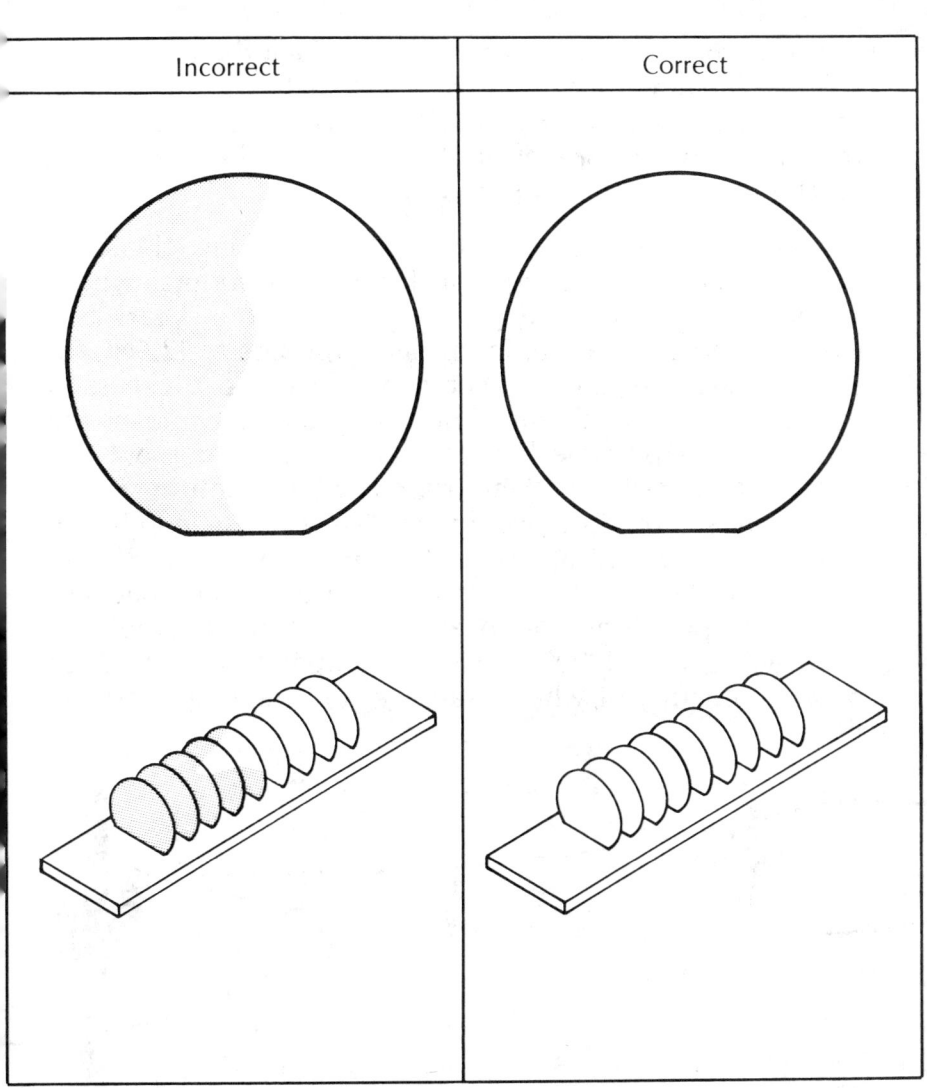

Oxide Growth

A number of factors control how thick the oxide layer will be. The amount of oxygen introduced into the furnace, the temperature at which the furnace is set, and the length of time the wafers are left in the furnace all have an effect on the thickness that is produced.

For example, it might be possible to achieve the same oxide thickness by putting a wafer into the furnace with a measured amount of oxygen at 500°C for three **years** as it would by putting a wafer in the same furnace at 1200°C for three **hours.** The goal is to get a specific oxide thickness in the most economical time period. This generally means the shortest time possible but not always. Why not put the wafer in at 2000°C for 10 minutes? The basic limiting factor is that our silicon wafer would melt like a popsicle in Death Valley! It has a melting point of approximately 1425°C and it becomes "rubbery" as this temperature is approached. For most operations, the maximum temperature would be no greater than 1300°C. Higher temperatures introduce other problems, of which warpage and melting head the list.

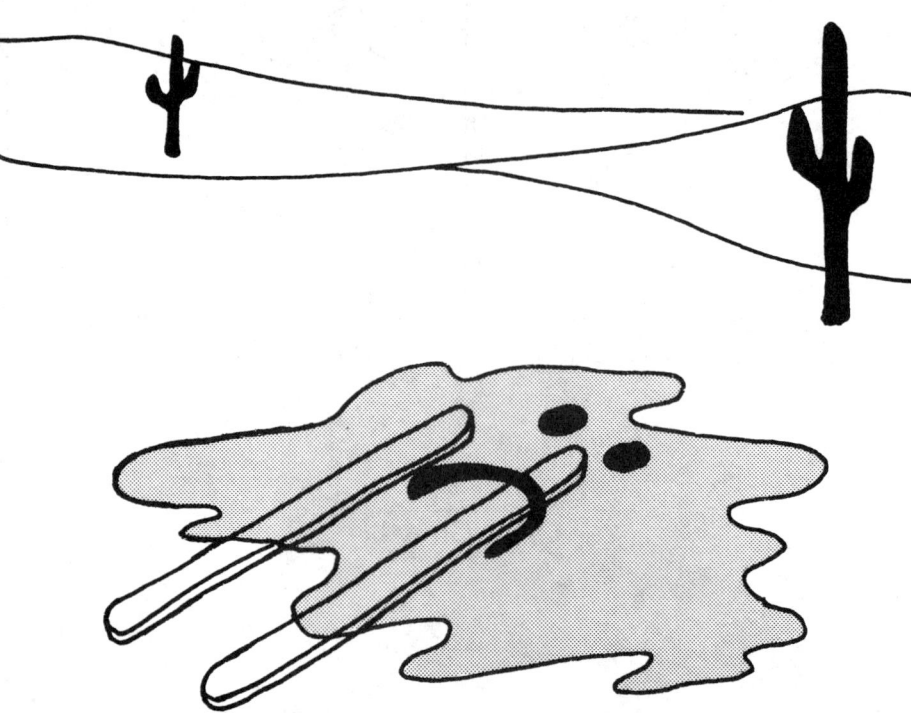

So, at initial oxidation — usually the first step in wafer fab — the object is to get the wafers oxidized quickly. However, at later oxidation steps other problems come into play. At these later oxidations not only is oxide being formed, but "diffusion" is also taking place. The recipes for these later oxidations and diffusions have been carefully calculated to be "the best of both worlds"; that is, the correct oxide thickness coupled with the correct diffusion. This combination will become more clear as it is discussed in the section on diffusion.

In the fab area, oxide is usually measured in terms of "angstrom units", or $\overset{\circ}{A}$, as it is shown in symbol form. An angstrom is a very precise form of measurement which is required in our manufacturing areas. It is equal to 1/100,000,000 centimeters (A dime is about 450 **million** angstroms thick!). It takes approximately 500 angstroms of oxide on a wafer before it can even be seen by the human eye. In most of the processing you will be exposed to, you will grow oxides from 500 to perhaps 10,000 angstroms in thickness. Some of these will be very crucial to the success of the product and will have tighter tolerance ranges (ie: 1275 $\pm$ 50 $\overset{\circ}{A}$) others may be less critical in terms of tolerance perhaps having a range of $\pm$ 500 $\overset{\circ}{A}$.

The normal question asked here is: "If at 500 angstroms you can barely see it, how can you measure it?" As oxide becomes thicker or thinner the color of the oxide on the surface of the wafer will change. For example, an oxide thickness of 500 angstroms, while barely visible, will have a brownish tint to it. If the thickness is increased to 1250 angstroms, it would appear blue in color. It would be relatively accurate to judge an oxide thickness by its color except that as the oxide becomes thicker or thinner the surface colors start to repeat themselves. While 1250 angstroms might appear blue, the color seen at 3100 or 5000 angstroms will appear blue also. There are other means of determining oxide thickness that are more accurate and reliable. Elipsometers are widely used in our industry and produce extremely accurate results. Your area will have a method in the process specification which will guide you in determining the thickness of the oxides you grow.

DIFFUSION

"Diffusion" is defined as the spreading of one substance through another. In wafer fab, this will consist of the spreading of "dopants" into the wafer. Dopants are chemical substances which, when added to the wafer in the correct amount, allow the circuit to regulate electrical current to make the circuit work properly. The dopants that we use are divided into two categories: "N" and "P" dopants. "N" dopants are Negative in charge, while "P" dopants are Positive in charge.

The following are N and P dopants which might be used:

N	P
Phosphorous	Boron
Antimony	Gallium
Arsenic	Aluminum

Within these groups they may be even further broken down by the chemicals that contain them:

N type Phosphorous	P type Boron
$POCL_3$	BBR_3
PCL_3	BCL_3
Phosphine	Diborane

A particular furnace operation that is called "N+ diffusion" on the run card means that it is an N type operation and that one of the "N" dopants will be used.

The engineering staff has selected one of these dopants because of some special properties that make its use better than one of the other N dopants.

The difference between N and P type dopants is as substantial as the difference between a negative or positive bank account. The difference between two different N types or two different P types is a little more subtle. If both are negative, why do we choose antimony over phosphorous for a certain operation? The selection of which is to be used is made early in the game, back when the circuit was designed.

It is agreed upon by both the engineer responsible for design and by the engineer responsible for the process. Once a decision has been made it is usually quite difficult to alter it. This is because the design of the circuit itself takes into consideration the properties of the dopant selected.

For example, in an N diffusion operation the engineers agreed that phosphorous would be used instead of arsenic or antimony because phosphorous has a different "diffusion rate" than either arsenic or antimony. That is, one type of dopant may take longer or shorter time to spread into the wafer than one of the others. This is most important to the design engineer in that he will consider the diffusion rate (spreading) of phosphorous when designing future steps in the process (subsequent diffusions). Once the circuit has been designed for using phosphorous, the use of any other dopant may have harmful effects on the product. The control of this diffusion is what the game of building electronic circuits is all about: too much diffusion is as bad as too little.

It would not be so complicated if we were making circuits with only one layer. In most processes we will have from five to fifteen diffusion layers in each wafer. In a process that has fifteen diffusions, the very first diffusion step is going to go into the furnace fourteen more times, meaning the spreading could continue each of those additional fourteen times. This must be accounted for by the designer.

Dopants, then, are chemicals which we add to the wafer. These dopants are then spread through the wafer when we "diffuse" it.

Below is an example of a single "dope and diffuse" cycle:

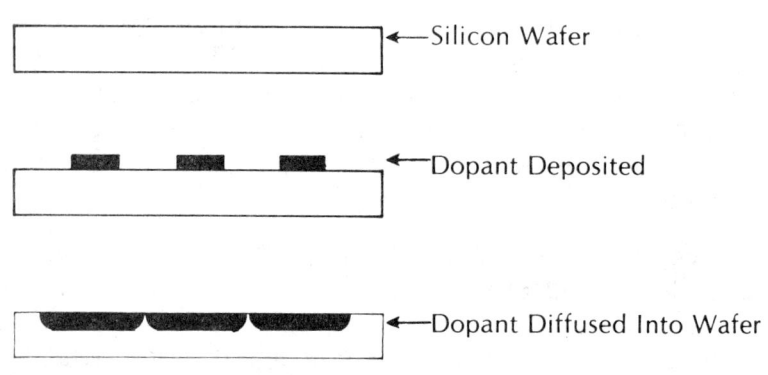

Dope & Diffuse Cycle

Diffusion is accomplished in much the same manner as in oxidation, in that a high temperature furnace is used. The dopant is driven into the wafer by the heat from the furnace. The depth (and width) of the diffusion is controlled by the time the wafer is in the furnace, and by the temperature of the furnace. To illustrate this, consider that a wafer with dopant on it is placed in an 1100°C furnace for three hours. A combination of time (three hours) and temperature (1100°C) will produce a diffusion of a certain depth and width into the wafer. If we left the time at three hours, but increased the temperature to 1200°C, the diffusion would be deeper and wider.

Now, let's look at combining two dopants on the same wafer, and two diffusions:

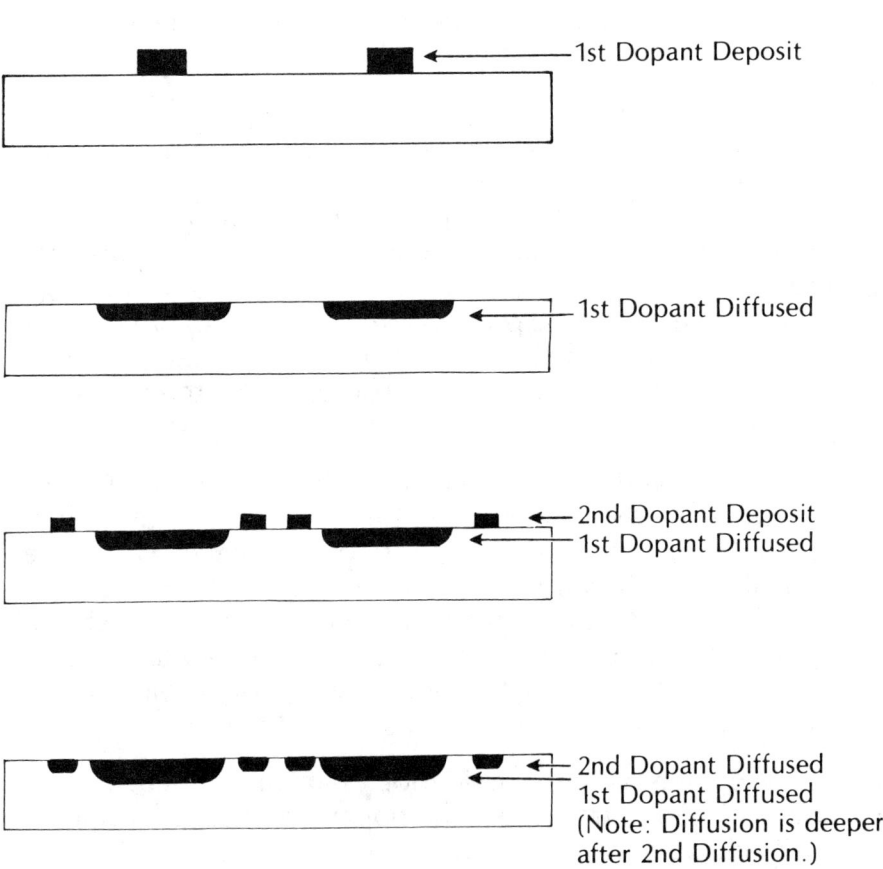

DEPOSITION

Although a combination of the various deposition techniques are used throughout the entire process, the most common practice is by placing wafers in a high temperature furnace and allowing a gaseous form of dopant to flow through the furnace. Some of this dopant lands on the wafers, while the excess is vented away. As in all matters of wafer processing, cleanliness and control are the key factors in obtaining successful results.

Not only must the correct type of dopant be used, as defined by the engineering staff, but also the amount that the wafers receive is of great importance. Very strict controls are set forth in the process specification (spec) which govern how much dopant is let into the furnace. There will be further items in the spec such as how fast or slow the run should be pushed into the furnace, or how fast or slow it should be pulled out, or what cleaning techniques are needed, or the positioning of the wafers in the furnace.

The dopant is deposited over all exposed surfaces of the wafer. At most operations this is really not desirable, but because of the miniscule size of the places where we want it to be, it is not practical to place it in only certain areas. So, the dopant is deposited over the entire wafer and then removed from the places where we don't want it. This is accomplished through a combination of oxidation and "masking".

We have seen how oxidation is accomplished. Part of how we can control where the dopant is to stay is done by removing some of this oxide.

Silicon dioxide (oxide) can be "etched" away by dipping the wafer into a chemical solution. This solution is hydrofluoric acid — usually called HF, or in a diluted form called oxide etch, 10 to 1, or other names. (NOTE: "Dry etch", an alternative to wet chemical etch, is described in the Glossary.)

An example of oxide etching is shown below:

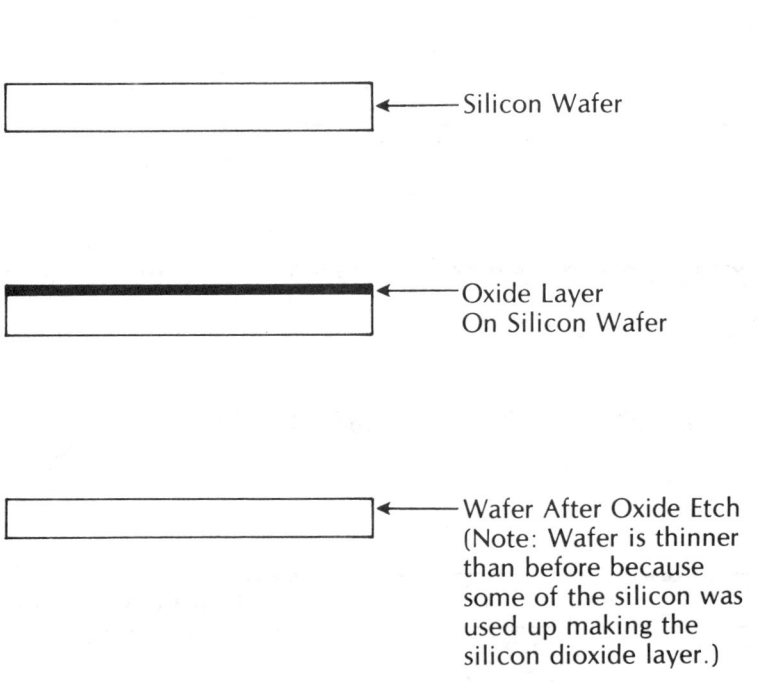

The problem is the dopant will be deposited across the whole wafer. We want to remove oxide from only certain areas — **the areas where we want the dopant to remain** — while leaving it in others as a protective covering. If the dopant is deposited on oxide, it will be removed when the oxide is removed by the etch solution. If the dopant is deposited on silicon (without an oxide cover) it will **not** be removed by the etch solution.

The following series of illustrations should help clarify:

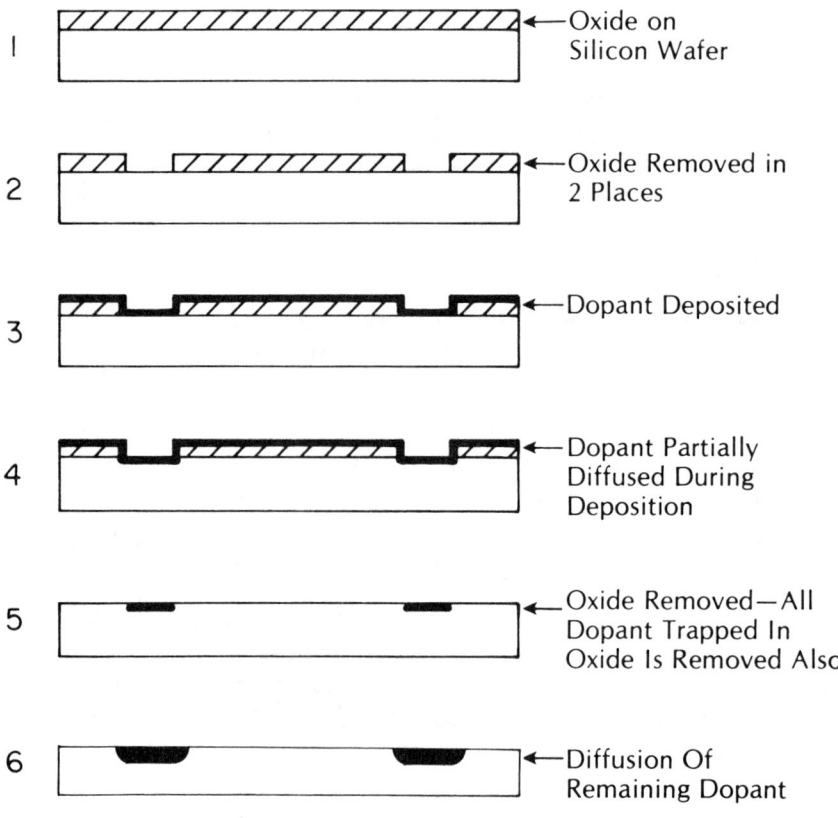

By removing the oxide from the desired places, we can control where the dopant will be on the wafer.

Now, there seems to be one big question left; how do we remove oxide from the **exact** spots desired? The answer: by "masking".

(NOTE: Alternative methods of deposition of dopants and thin films are discussed after the section on "Assembly and Final Test." See "Deposition...revisited" starting on page 65.)

MASKING

Masking, which means to conceal or hide an object, is carried out in many forms. In painting an automobile, we mask the chrome bumper when painting near it so that paint does not get on the chrome surface. In art, wax is used to cover areas of glass while etching a pattern in the exposed area of the glass.

In semiconductor processing, masking is used in the same manner — to protect one area of the wafer while working on another. This process in semiconductors is called photolithography, photo-masking or "masking".

If you have ever been in or seen a darkroom, you know that if you turn on a light while film was being processed the film would be ruined. The same is true when processing wafers in the masking area. In would be very impractical however, for 30 to 100 people to be bumping into each other all day long in a darkened room, so a compromise has been used — yellow light. The yellow light allows everyone to see and at the same time it is safe for the type of processing that takes place in the masking area.

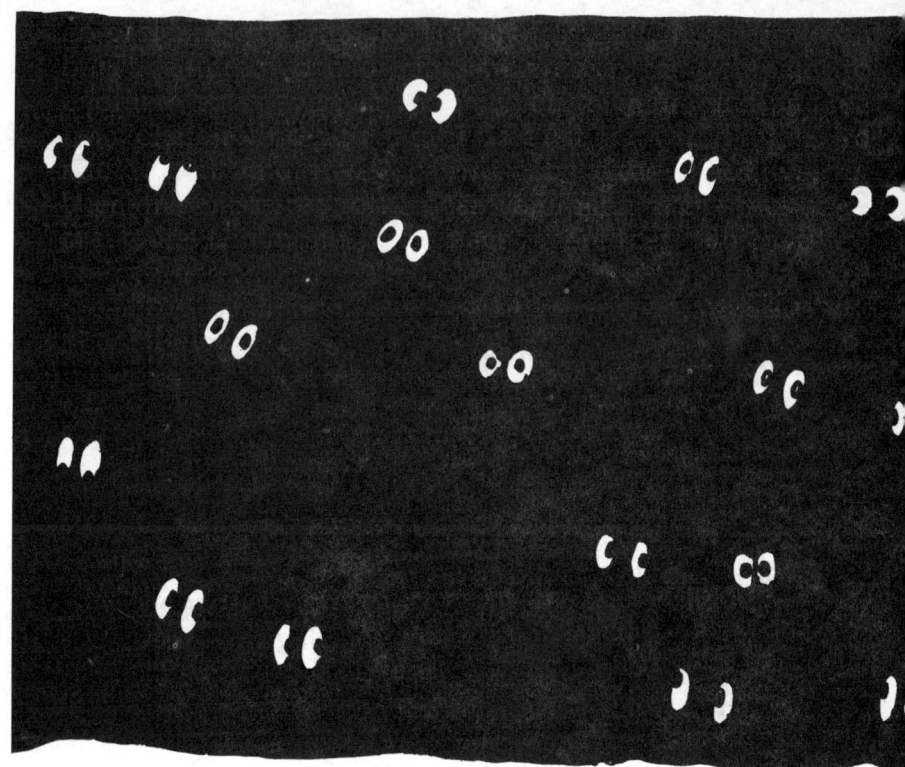

Now that we're not in the dark about why the room is yellow, let's work through the masking process.

A chemical, called "photoresist", is used to start the masking process. It can be compared to the film we use in an ordinary camera. Photoresist is white light (ultraviolet) sensitive, meaning that it will be exposed in white light. It comes in liquid form and its thickness, called viscosity, can be varied from one production area in the company to another. It can even vary within your own production area from one mask step to another (ie., 1st mask, 2nd mask, 3rd mask, etc.). Photoresist can also be either "positive" or "negative". Basically, the positive process is the reverse of the more commonly used negative process which is explained here.

The manner in which the photoresist, or "resist", as it is commonly called, is applied can also vary. At one time not that long ago, it was squirted manually onto the oxidized wafer with a syringe and then spread very thinly and evenly over the wafer surface as it was spun rapidly by a machine. This was done one wafer at a time by an operator, and of course, was very time consuming. Later, a machine was developed that would automatically feed one wafer at a time from a stack of wafers onto the spinning mechanism, automatically dispense the resist, spin the wafer, and then send it off the other side of the machine while the next wafer was being fed up onto the "spinner". The advantages were great because one person could spin many times the number of wafers as before. Plus, if the machine worked properly, that person could also be working on some other aspect of the masking operation.

Further developments have been made that perhaps your area now employs — such as systems that not only do the operation which loads — spins — unloads automatically, but even carries the wafer through many more subsequent operations in the masking process. Whatever method or combination of methods your area employs, it is important that the spinning of the resist produces a clean, even-thickness layer of the correct type of resist.

One of the properties of resist is that it is quite soft and sticky. Because of this, and to help remove solvents from the resist, the resist-spun wafer is "baked" in an oven. The time the wafer is baked and at what temperature is critical to the forthcoming process steps and again, it is necessary to follow the process spec to the letter.

After the baking has been concluded the actual masking takes place. This is done by a process of "alignment" and "exposure". The machine used for aligning can vary in its size, overall appearance, and its method of operation. However, the function that all these machines — called alignment "jigs", alignment "tools", "aligners", or various other names — is the same. The wafer is placed onto this machine and a **specific** patterned "mask plate" is placed over the wafer. The wafer is then "aligned" properly with the mask plate, and then exposed through the action of the shutter of the machine opening to allow ultraviolet light to hit the unmasked portion of the wafer.

Got it? All these new words hit pretty fast. Let's examine in more detail what is happening. The alignment jig is a very complex machine with all kinds of buttons, knobs, pressure gauges, and other do-dads. It even sits on shock pads (to absorb vibrations) but basically, it is like a camera for the masking operation. In very simple form, it operates as follows:

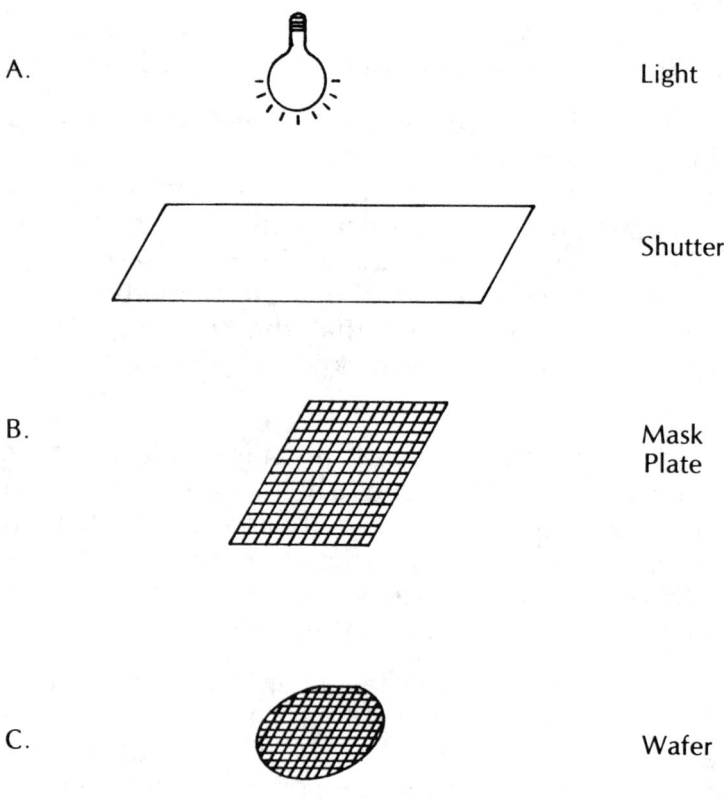

The light (A) from above shines through a mask plate (B) and exposes the resist on the wafer (C).

An alignment jig can be fully automatic, semi-automatic, or manual. It might also be a "projection", "proximity", "E-Beam", or stepper". (These are described in the glossary). In the manually operated machine the operator performs a series of operations which make up the alignment job. Assuming the correct mask plate has been selected to coincide with the wafers which are to be aligned, the mask plate is placed into the space provided for it on the machine. Next, a wafer which has already been spun and baked, is positioned below the mask plate on the "wafer chuck" (the flat or notched portion of the wafer is used as a reference point in placing the wafer on the chuck). Using the microscope attached to the jig, the operator looks down through the clear areas of the mask plate to the wafer. The wafer below the mask plate is then rotated (on the chuck) until the masking "alignment marks" of the wafer and the mask plate are in the same position. When these marks coincide, the shutter between the light and the mask plate is opened. The light rays pass down through the **clear areas** of the mask plate and exposes the resist on the wafer below the clear areas. The darkened portion of the mask plate **does not** allow the light to pass through it so the resist on the wafer directly below the darkened portions is **not** exposed.

Like so:

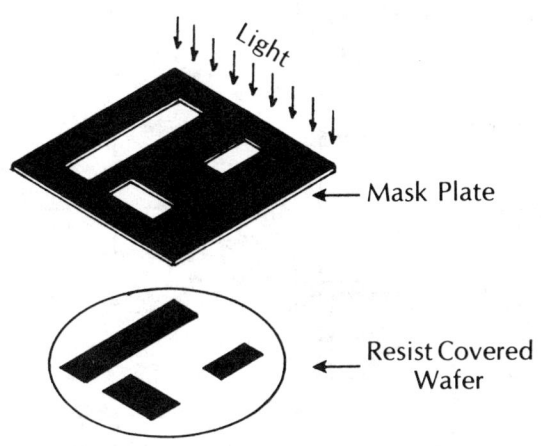

← Mask Plate

← Resist Covered Wafer

The resist is exposed only where the light passes through clear areas of the mask plate.

The alignment mark, usually made in the shape of boxes or crosses, is used as a guide for aligning the wafer to the mask plate properly. It generally serves no electrical function on the circuit but is extremely important in getting all the mask steps lined up correctly.

Alignment marks, located on each "die", are representative of the circuits on the wafer. That is, if the alignment marks are aligned properly, then all other portions of the circuit will be line up properly.

A series of alignment marks on mask steps 1 thru 4 might look like this:

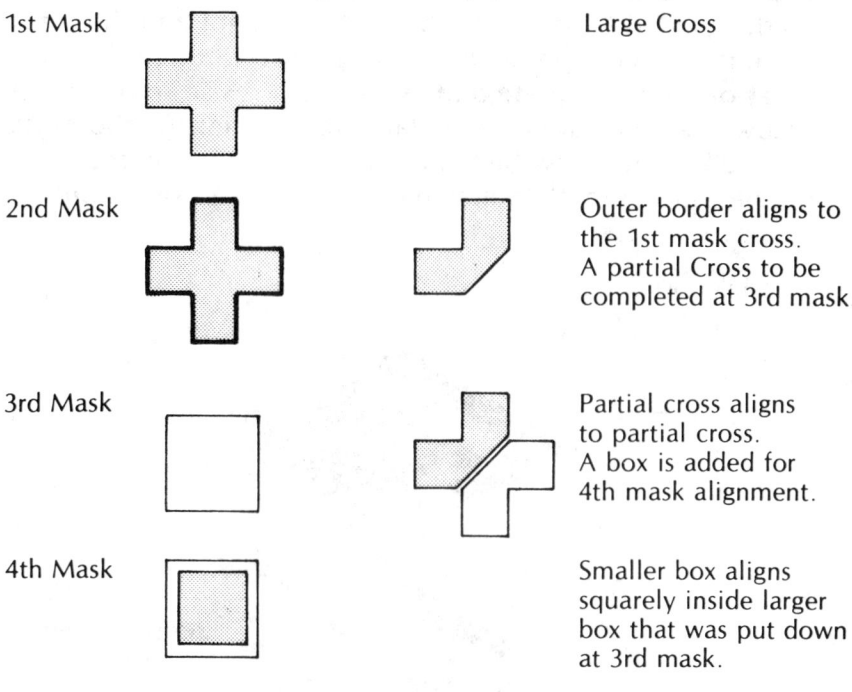

Examples of incorrect alignment or "misalignments" would look like this:

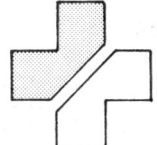
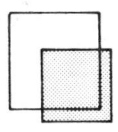

2nd mask dark border misaligned to 1st mask cross

3rd mask half cross misaligned to 2nd mask half cross

Smaller 4th mask box misaligned to larger 3rd mask box

If you examine any mask plate you will see that each die on it is the same except for possibly a few test die. This is because we are not building just one circuit per wafer. Each die, after stacking all the mask layers up will make the circuit. If the die are small, there may be thousands of circuits on each wafer. On the other hand, larger size die will make less than a hundred on a wafer.

The circuit will be built on the wafer one layer at a time.

One die, when enlarged, may look like the following (remember each die on a mask plate will look exactly the same as all others — minus the test die).

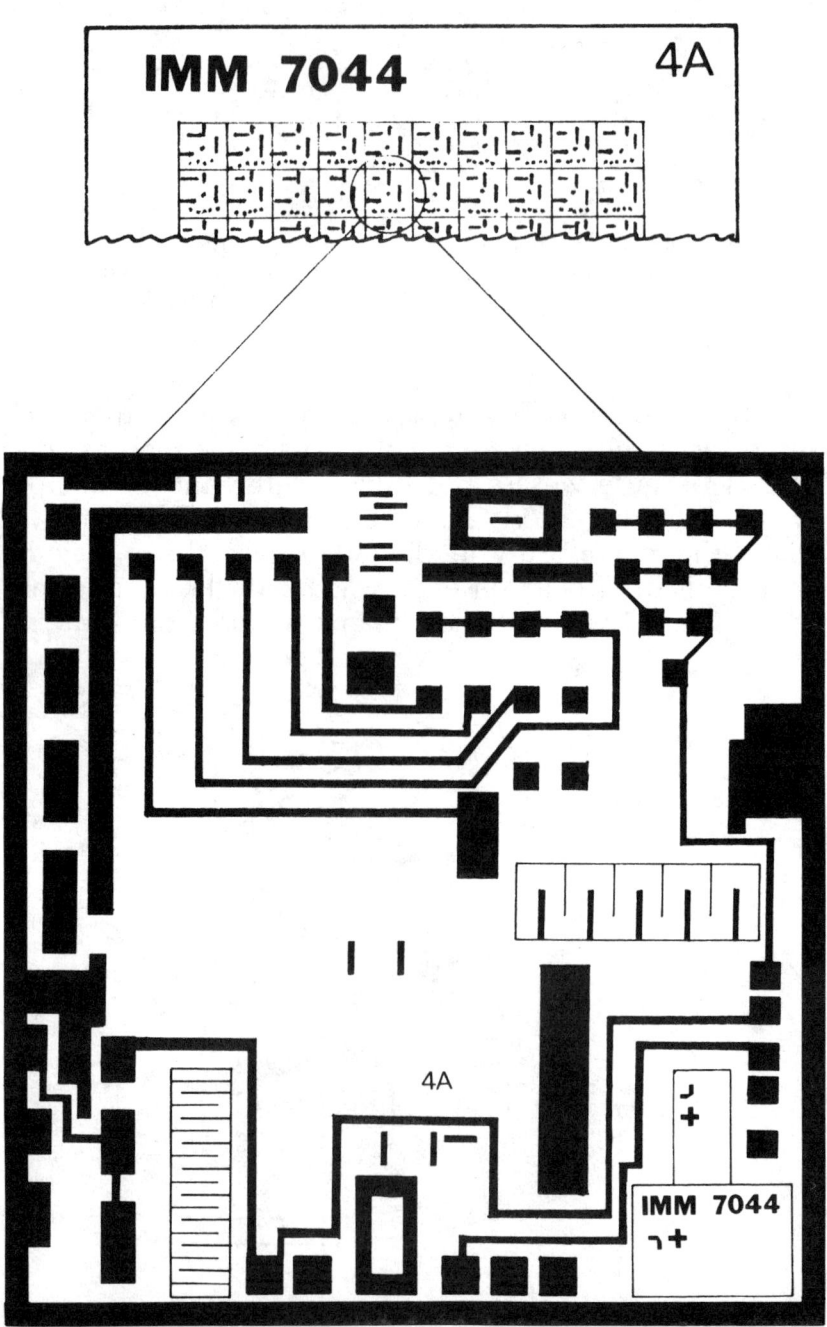

A complete "mask set" (all the layers that are to be masked in the process, ie. 1st, 2nd, 3rd, 4th, etc.) will look very similiar in that: 1. they are the same size, 2. they will have the same product number (each product will have its own set), 3. the alignment marks from one mask to another within the set will be in coincident places.

While similiar, they are not the same. The circuit is being built one layer at a time, one on top of another.

Think about it like a SUPER Dagwood Bumstead sandwich that you like to eat when made in one exact precise way. The salami, the mayonnaise, the lettuce, the pickles and tomatoes must go on the bread in the exact manner for it to be **perfect.** If you omit the tomatoes it is no longer perfect. Or if you put the pickles in the wrong spot, it is no longer perfect (misaligned pickles??? — this is getting out of hand!!).

The same goes for the wafers. If you omit 3rd mask, or if you misalign 3rd mask, the wafer is no longer perfect.

The big difference is that you'll probably still enjoy your sandwich, even though it is not **perfect.** Whereas, **no** one will enjoy your wafers because they won't work and will end up in the garbage. (NOTE: there is usually some tolerance allowed for alignment, your process spec or supervisor can tell you how much is allowable. There is no tolerance for an omitted mask and the wafer or wafers will be rejected).

RATZ! EVERYONE KNOWS THE MAYONNAISE IS SUPPOSED TO GO NEXT TO THE LETTUCE!

Looking back at the enlarged die, you can see how complex and miniscule things are. A microscope is essential when aligning or inspecting a die or wafer. The microscope will enable us to see things as many as 1000 times their actual size — an invaluable tool in doing quality work.

After the wafer has been aligned and exposed, the next step, like in photography, is developing.

In developing a wafer, a machine usually very much like the spinning machine is used. Unlike spin, where we put the photoresist on, developing takes it off — almost. Actually, it will only take the resist off the wafer in the areas where the light was **not** allowed to pass through the mask plate (the unexposed portions of the wafer).

The developing is done by chemicals which are sprayed down onto the wafer. This spray washes away the non-exposed resist while the exposed or "polymerized" resist remains.

The next step in the masking process has historically been crucial to attaining high quality products. It is called develop check or develop inspect. The function here is to inspect the accuracy or quality of the work performed on the wafer through spin, align/expose, and develop. A microscope is used and usually each wafer is inspected in a few places — looking at each die on a wafer would take much too long. The inspector looks for such things as:

1. Proper develop
2. Mask pattern exposed on the wafer
3. Proper alignment
4. Correct mask used
5. Other items as outlined in the spec.

Wafers which have thus far been completed properly — those which pass the requirements of develop inspect — are sent on to be baked again. Those which fail inspection can have the remaining resist removed and recycled (re-worked) through the operations again.

The wafers which pass develop inspect are baked to

remove any chemicals which may have remained on them at develop, and then sent on to the etch area.

In addition to the properties of resist which were already mentioned, another important property is that resist is impervious to oxide etch. So, it follows that if the resist will not be etched away, then it protects the oxide under that resist also. Where the resist has been developed or washed away the etch will remove the uncovered oxide. The following illustrations show how this works:

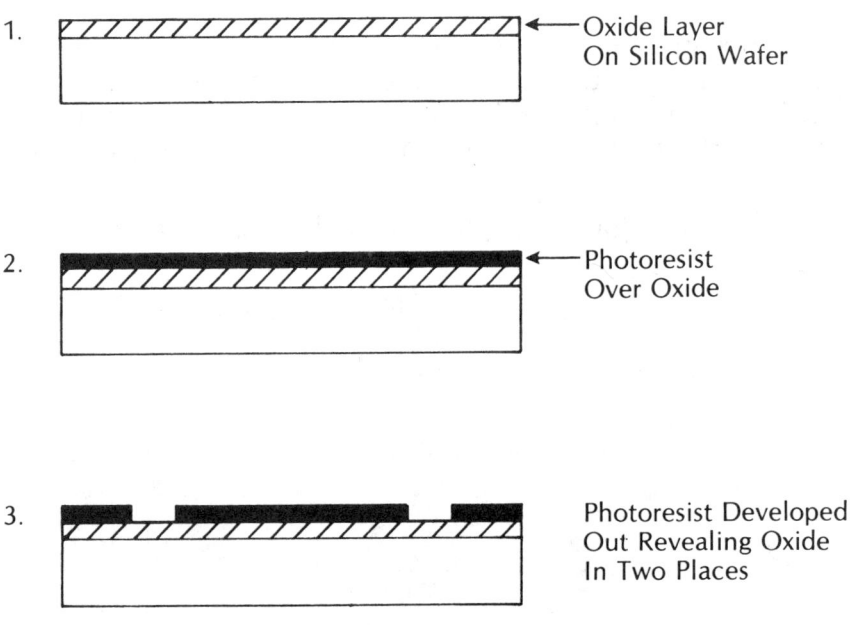

If we now put a wafer which looks like sketch number 3. in the oxide etch solution, the acid can only etch away the oxide where the acid comes in contact with the oxide:

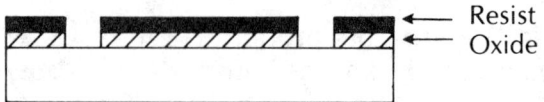
← Resist
← Oxide

By putting the wafer into another chemical bath which removes all the remaining resist, we will end up with a wafer like this:

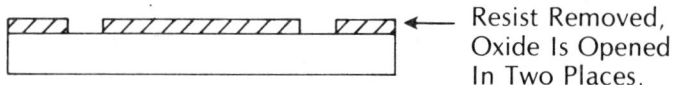
← Resist Removed, Oxide Is Opened In Two Places.

Because of this process we are able to choose the exact places on the wafer where we want oxide to remain. This is most significant in that we now can put the dopants to be diffused into the wafer in **only** those places where we want them.

So, the flow through a typical masking area would look like this:

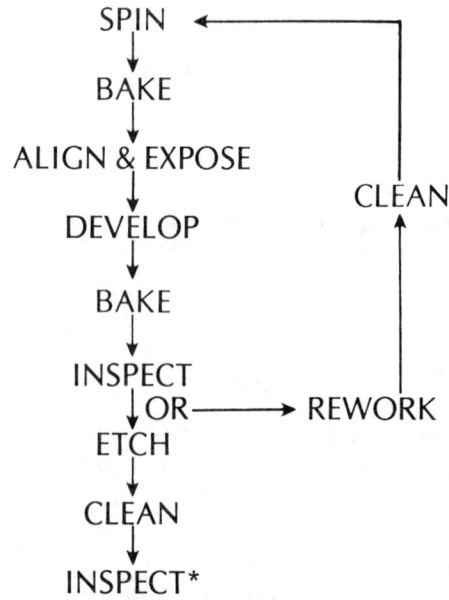

With few exceptions this flow is repeated for each masking step in the process (1st mask, 2nd mask, etc.)

Upon completion in masking the wafers usually travel back into the diffusion area for another slightly different cycle of deposition, diffusion and oxide growth. They then return to masking for the next mask step, and so on until all steps of the process are complete.

The last inspection guarantees that the etch was complete and that all the resist was removed from the wafer before it is passed on to the next operation.

PROTECTIVE CIRCUIT COATING

The final manufacturing fab operation to be completed when making a wafer is putting a protective seal over the circuit. This operation is also called by many names such as vapox, glass, silicon dioxide, etc., and is generally referred to as final "passivation".

Because of the relatively low melting point of the aluminum on the wafer (approximately 660°C), this operation is not done in a typical furnace. It is usually done in a CVD system (chemical vapor deposition) where a thin coating of oxide material is deposited over the entire surface of the wafer. The wafer is then sent to a final masking step which opens windows or pads that will be used later when bonding the circuit (see the section on assembly for bonding).

ELECTRICAL TEST

The final step in wafer form of integrated circuit manufacturing is testing. At electrical test ("die sort", "wafer sort", "wafer probe," etc.) each circuit or die is tested for its ability to perform the operations for which it was designed.

The first step in this process is to determine which electrical characteristics will be tested. The test engineer and circuit design engineer spend many hours writing a computer program which will specify the functions to be tested that will guarantee the circuits acceptability. Then, a "probe card" is built which will be used to physically contact the bonding pads of each circuit and feed information into the computer which will determine if the circuit is accepted or rejected.

A machine, called a wafer sorter, is hooked to the computer. The probe card and one wafer are placed on the sorter. When properly aligned (the probes of the probe card to the bonding pads of the wafer) and proper contact has been made, the testing starts.

The sorter indexes from one die to the next until the entire wafer has been tested. Testing one wafer can take from less than one minute to over an hour, depending upon the complexities of the circuit and the test program.

As each die is tested, the computer records certain information about it. If a die is not acceptable, that is, if it fails any one or more of the tests, a small droplet of ink is automatically placed on that die so that when the wafer is separated into individual die the bad or "inked" die can be discarded.

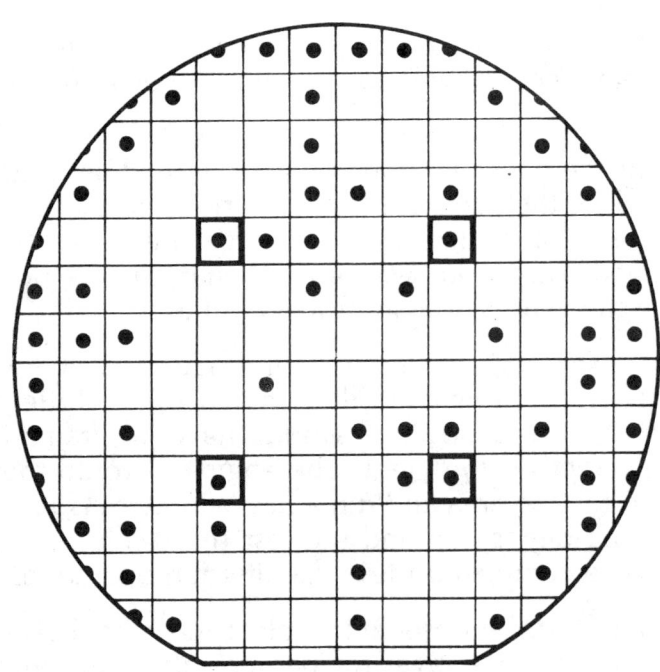

"INKED" Wafer After Sort

When a wafer has been completely tested, the computer will print out valuable information about it. This information will be used by many different people throughout the company.

For example, knowing the yield (the number of good die compared to the number of die possible on the wafer) will be of assistance to production because after several wafers, or several runs of wafers of a certain type, the number of wafers it will take to meet the schedule on that product can be determined. This is how the manufacturing manager knows how many wafers to start each week.

The computerized yield information also tells production control how many wafers to buy to meet the start schedule, or how many parts will be sent to assembly.

The marketing people use this information to help in determining how many circuits will be available to sell and what the price should be.

The production and design engineers can check to see if this yield meets their (and their bosses) expectations for this product.

If the yield is much lower than it should be, a further evaluation of the computer's information can tell what the major areas of failing were, and generally, it can be determined what went wrong and what can be done to improve the wafers now in process in fab.

For example, from the information gathered at electrical test, it might be determined that a little more dopant at 2nd deposition would have a dramatically positive effect on the yield. The engineer would then go back into the fab area and run a few sample runs to a new spec at 2nd deposition and the test the results when the wafers have completed the remaining fab operations.

It may also be observed at electrical test that a new mask design for one layer (or perhaps all) could improve the yield results.

There are hundreds of possibilities for yield improvement that can come as a direct result of the data collected in electrical test, and these results can affect many people throughout the company.

ASSEMBLY & FINAL TEST

Upon completion of testing, the circuits need to be built into packages which the customers can effectively use.

The first step in assembly is to render individual die from a wafer full of many die. This can be accomplished by diamond scribing, sawing, or laser scribing — again, each company, and indeed, each division within a company, may employ a different method.

Basically, all three methods will cut a groove down into the wafer between each die (in the scribe lines). When all the scribe lines running from top to bottom have been cut, the wafer will be rotated 90°, and then cut again so that all of the boundaries of the wafer have been cut. The wafer is then "rolled" so that the die will break free and become individual die.

Die Separated From Wafer

The inked die (the rejects) will be separated and

discarded leaving a group of good functioning die.

Using vacuum-tip "wands", the good die are put into plastic trays and visually inspected. Using microscopes, this inspection is aimed at finding die which are dirty, scratched, chipped, or inked (some of the inked die may not have been pulled at the last step). There may also be specific items which a customer will not accept as good even though the die function properly. Discolored bonding pads, slightly misaligned mask layers, ragged edges on a mask layer are but a few of these special requirements by customers.

The die which pass this inspection which are to be assembled will be sent on to the "die attach" station. Here the die is attached to the package housing, called a header. This is usually done by heating the header and "scrubbing" the back of the die into molten gold on the header. Sometimes an additional piece of gold, called preform, is used, this usually occurring with larger die or for use in ceramic packages.

The packages are made of metal, ceramic or plastic, and come in a variety of shapes and sizes.

Some packages may look like this:

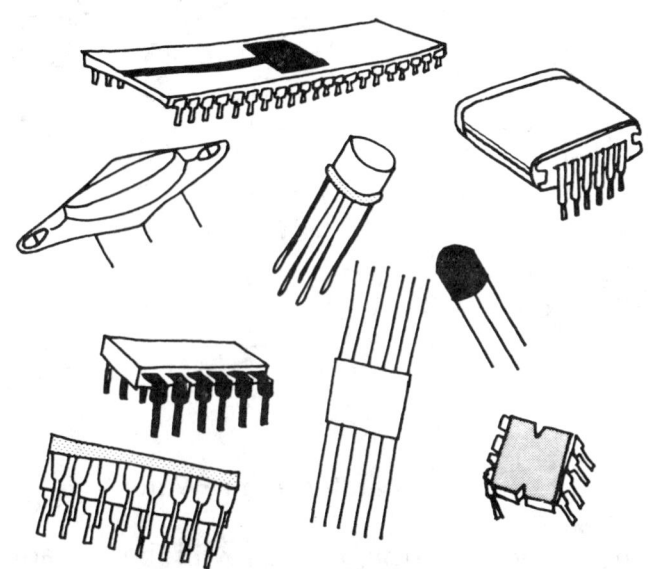

Once the die is attached to the package and another cleaning step has been accomplished, bonding takes place.

Bonding is the attaching of very thin wires which connect the bonding pads of the circuit to the leads of the package. For each product type there is a bonding diagram which shows which package leads go to which bonding pads. The package leads will be used as the connecting posts which will be inserted into the sockets of a television, or a watch, or a computer, etc. If the bonding pad is connected to the wrong package lead the circuit will not function.

The machines used to bond the circuit will vary, but generally they will use either gold or aluminium wire approximately one mil in diameter (about one-fourth as thick as a spider web). Because of the size of these wires a microscope is used so that these intricate bonds can be made accurately.

The number of bonds made per each package will also vary. There can be from as few as two up to as many as forty-eight or more per package.

When the bonding has been completed, another cleaning and inspection is usually done as well as a bond pull test; all these steps help guarantee that the bonding has been done properly and that the product will "hold up" when it is used by the customer.

The packages are then sealed and checked for air tightness and water tightness and then shipped to the final test area.

FINAL TEST

Integrated circuits which have spent a great deal of time traveling through the manufacturing process are now ready for one final evaluation prior to being sent to customers who will plug them into the systems for which they were designed.

In final test, packaged parts are run back through the computer, much like the test which was used at electrical test when the die were still in wafer form.

Some packaged units will fail at final test even though they were "good" at electrical test. This is because contamination, broken or missing wires, incorrect bonding, or a variety of other reasons caused a problem.

It is also possible that some units will work better than others ("faster" or function at a lower range etc.). Many times a customer will exist for both the better units as well as the slightly lower quality parts. In this case the units will be classified into which group they belong and will be marked and sold accordingly.

If not previously marked with the product type, the units will be sent to the marking area and then packaged into boxes for shipment to the customer for his use in the thousands of products in which IC's are used.

DEPOSITION...REVISITED

ION IMPLANT

As with most of the machinery and equipment used in the manufacturing of microelectronic circuits, the ion implanter looks about as easy to operate as flying a Boeing 747. Actually, the operation and function is fairly easy. What we are trying to accomplish at "implant" is usually no different than what we do at a deposition step in a furnace.

As in most deposition steps, the wafer is "masked" off so that the dopant which we wish to implant can only go into selected parts of the wafer. And, like a furnace deposition step, the dopant may be diffused into the wafer at a later step. The reason for using this large, expensive, and complicated looking machine is for controlling the amount of dopant needed. The amount of dopant desired is so small that a furnace operation could not provide the accuracy required.

The machine works by segregating certain ions in the dopant gas (by an electro-magnet) and then accelerating the ions in a long beam which blasts into the wafer. Again, like a deposition cycle done in a furnace, the ions will penetrate the bare silicon but will not penetrate where oxide (or other blocking material) is left on the wafer.

"EENIE MEANIE MINEY MOE......"

EVAPORATION

In processing different types of integrated circuits several different evaporations are used. Gold, aluminum, and silicon-aluminum are some of the more commonly used. All of these materials are put on the wafer in much the same manner. Aluminum is the most common, so it will be used to illustrate.

Evaporation ("evap") means vaporizing a chunk of very pure metal and letting the vapor deposit on the wafer.

This occurs when a heat source (electron beam or other sources) is directed at a crucible full of pure aluminum. When the beam hits the aluminum and is hot enough, the aluminum will vaporize. This can be visualized and more readily understood if you imagine it like a teakettle full of water being heated on a stove. When the temperature of the stove is hot enough the water in the kettle starts to vaporize, putting off steam. If you were to put an object above the steam coming out of the kettle the vapor would condense (return to water) as it hit the object.

In aluminum evaporation, the wafer becomes the object above the heated and vaporized aluminum. The vaporized aluminum then condenses in a very thin and uniform layer on the wafer.

There is, naturally, a special piece of equipment used for this evaporation process; it is, oddly enough, called an evaporator! It is very important that the equipment system and the wafers both be very clean, because the evaporation takes place under vacuum in the system. When a vacuum is created, all air, water, and particles are pumped and filtered out of the evaporation chamber; this way a clean aluminum deposit can be made.

The purpose of aluminum evap is to provide the "wiring" for the circuits on the wafer. At a later masking step part of the aluminum will be etched away, leaving only connecting spots and thin lines of aluminum which act as the wiring for the circuits.

SPUTTERING

Sputtering, as a deposition method, is accomplished by placing the wafers in a vacuum chamber and pumping it down. Then, by precise control of electrical current, a small amount of "ionized gas" bombards a "target" (the material we want to deposit). This results in some of the target material being dislodged and re-distributed in a thin film onto the wafers.

Some materials that are typically "sputtered" include targets of aluminum — an alternative to aluminum evaporation to make the circuits' metal interconnects, silicon nitride — generally used for a "passivation" layer in protecting the circuit from corrosion and moisture seeping into it. Targets made of other materials including platinum, molybdenum, tungsten, etc., or combinations of these may be used.

CHEMICAL VAPOR DEPOSITION (CVD)

Depending upon the requirements of the process or the inclination of the engineering group, CVD's can be accomplished in a variety of ways. The options here are indeed many. CVD's can be done in vertical or horizontal systems. They can be at low pressure or at atmospheric pressure, high temperature or low, plasma enhanced, radiant heated or even use ultraviolet light as their energy source.

Some of the films deposited include silicon dioxide — used generally as a "passivation" layer, polysilicon ("poly") — primarily used in making silicon gate MOS products, silicon nitride — again, primarily used as a passivation layer, epitaxial films — vaporized silicon deposited on the wafers wherein the "epi" film conforms to the exact crystalline structure of the wafer.

Perhaps the most common CVD is low pressure (LPCVD). This has generally taken place in a system which looks much like a standard diffusion tube but is sealed at both ends such that a vacuum can be pulled (low pressure). The lower pressure allows for more uniform and higher quality films to be formed. Even higher quality films can be produced in a system that employs a vertical loading method for the wafers by reducing the possibility of particles falling onto the wafer surface.

ABC's of IC's
(Glossary)

A/D Converter (analog/digital) A circuit which changes information presented from analog to digital form. (also can be D/A)

Air Bearing or Air Track Method of conveying wafers from one work operation to another on various machines used through the manufacturing process.

Align In masking, the "lining up" of two or more layers so that the components of one layer are compatible with the components of the other layer.

Alignment Aid or Mark Boxes, lines, crosses, etc., drawn on each mask layer to aid the operator or automated alignment machine in getting the pattern of the mask plate placed correctly on the wafer.

Alignment Jig The machine used in masking to align the pattern of the mask plate to the pattern on the wafer and to expose the photoresist. Also called alignment tool, or aligner.

Alloy The combining of two or more elements on a wafer to attain better adhesion. Commonly, the furnace process after metal masking for the purpose of getting the aluminum or aluminum-silicon to remain bonded to the wafer.

ALU Arithmetic Logic Unit.

Aluminum A conductive metal used in wafer fab to connect the various parts of the circuit. Aluminum wire is also sometimes used in bonding to connect the pads of the circuit to the leads of the package.

Angstrom A unit of measure to gauge thickness of various layers of material (oxide, aluminum, etc.) An angstrom is equal to 1/100,000,000 centimeter. The symbol $\overset{\circ}{A}$.

Anneal Generally, a furnace operation designed to repair damage to wafer structure caused by ion implant. Can also be accomplished by more sophisticated means via laser, incoherent light, etc.

Antimony A chemical element used as an N-type dopant in crystal growing and various deposition processes. Symbol: Sb.

AOQ Average Outgoing Quality.

Applications A sector of engineering that deals with which type of circuitry is best for a variety of customer uses.

AQL Acceptable Quality Level.

Architecture The design or method chosen to draw out a circuit.

Argon Inert gas used as a carrier gas for other chemicals in deposition systems.

Array An orderly grouping or arrangement of items. A Gate Array, for instance, will have an orderly grouping of gates that can be connected in a variety of ways during the later fab operations, as opposed to other circuit types that are predetermined.

Arsenic A chemical element used as an N-type dopant in crystal growing and various doping and deposition processes. Symbol: As.

Art Work The large drawings of the various layers of a circuit. It is used to make the master mask for each layer.

ATE Automatic Test Equipment.

Assembly The manufacturing process which converts circuits in wafer form into finished packaged parts.

Barrel/Etch/Strip Dry etch technique which uses caustics to remove oxide (etch) or photoresist (strip) from the wafer.

BiMOS The combining of Bipolar and MOS design and construction on the same wafer.

Bipolar Literally, having two poles. Generally refers to the process used in manufacturing of IC's that have both N and P type carriers present. (i.e., ECL, SCHOTTKY, TTL)

BIT Binary Digit.

Boat See **Sled.**

BOE Buffered oxide etch.

Bonding The connecting of a wire from the package leads to the pads (bonding pads) of the circuit.

Bonding Diagram A map which shows the bonding operator which package lead should be connected to which pad.

Boron A non-metallic element used as a P-type dopant in various fabrication processes (depositions, crystal growing, epitaxy, ion implant, etc.) Symbol: B.

Byte Generally 8 bits consecutively lined together as a unit.

C-Centigrade The unit of temperature measurement used in integrated circuit manufacturing. In degrees Centigrade, 0 = freezing and 100 is the boiling point of water. Symbol: C°.

CAD Computer Aided Design.

CAE Computer Aided Engineering.

CAM Computer Aided Manufacturing.

Cassette Slotted vessel which will hold wafers for cleaning, transporting or processing through the entire manufacturing process.

CAT Computer Aided Test

CCD Charged Coupled Device.

CD See **Critical Dimension.**

Cerdip Ceramic Dual-In-Line Package

Charge In crystal growing the cylindrical shaped piece of pure silicon which is melted and then grown into a single structure crystal.

Chip One integrated circuit. Also called Die, Circuit, Unit, etc. Many chips are made on one wafer then separated and packaged individually.

Class Classification of finished products into various categories of acceptability. (i.e., some products will work faster than others).

Clean Room Manufacturing areas where particle contamination, if not maintained at very low levels, could have disastrous effects on yields.

CML Current-Mode Logic.

CMOS Complimentary Metal Oxide Semiconductor.

Coat Masking operation where photoresist is applied to a wafer and rotated at high speed so that the resist coats the entire wafer with a uniform film (also called **SPIN**).

Comparator Optical measurement equipment used to check various physical dimensions of the wafer (i.e., critical dimensions).

Conductor A material which can conduct or pass electrical current. (Copper is a good conductor, wood is not.)

Contamination *Any* foreign matter, which in the presence of wafers, equipment, tools etc., causes product or process failure.

CPU Central Processing Unit.

Critical Dimension In masking, areas of the circuit which when measured for width must meet certain specifications to be acceptable. Also called CD.

CRT Cathode-Ray Tube.

Crucible A heat resistant cup used for holding molten material (i.e., in crystal growing a quartz crucible is used to hold molten silicon during the growing process).

Crystal The single-structured silicon ingot from which wafers will be sliced. Also symbolized as "X-TL".

Crystal Growing The process of transforming poly-crystalline silicon into single or mono structured silicon crystal.

Current The flow or rate of flow which electricity passes through material.

CVD Chemical Vapor Deposition. A process that chemically isolates and deposits a specific material on a wafer.

C-V Plot Current-Voltage tests performed in wafer fab which give a relative measure of certain types of contamination that when present lead to process breakdown.

Czochralski or **"CZ"** Most common method for growing the silicon ingot which will be sliced into wafers for processing.

D/A See A/D.

Dehydration Bake A masking process where the wafers are baked at low temperature prior to processing to remove surface moisture so to result in better adhesion of the photoresist.

Deionized water Water which has had all atomically charged particles removed. Commonly called "D I" water, it is used throughout the entire manufacturing process in cleaning, diluting of chemicals, and as an ingredient in the oxidation process.

Deposition The depositing or laying down of various chemicals on wafers. It most generally refers to the introduction of dopant to the wafers in a high temperature furnace, CVD's, sputtering, implant or evaporations.

Design The method or rules that are used in determining the function and layout of a circuit.

Develop The chemicals sprayed on photoresist and the removing of unexposed resist from the wafer during the masking process.

Device An IC product, it is usually represented by a code name. (i.e., an LM 109 is a device made by a Linear product group).

Die See **Chip.**

Die Attach The process of affixing the die to the package in which it will be used.

DIP Dual-Inline-Package; a common package used in IC manufacturing.

Discrete or **Discrete Device** Opposite of an integrated circuit (different devices on the same chip), these devices could be all transistors or all diodes but not a combination of different devices.

D I Water See **Deionized Water.**

Dopant Chemical "impurities" used to regulate the current flow in integrated circuits. Usually put on the wafer via furnaces, implants or CVD systems and later diffused further into the wafer.

DRAM Dynamic Random Access Memory.

Dry Etch Generally used in place of the acid bathing technique to produce more uniform pattern definition, particularly with smaller geometries, as is necessary for VLSI processing.

DSW Direct-Step-on-the-Wafer. A masking process in which the alignment machine uses the reticle instead of a photomask for placing the circuit pattern on the wafer (also called STEPPER).

DTL Diode-Transistor Logic.

DUT Device Under Test.

DUV Deep Ultraviolet.

Dynamic Ram See **DRAM.**

EAROM Electrically Alterable Read-Only Memory.

E-Beam See **Electron Beam Lithography.**

ECL Emitter-Coupled Logic.

EDP Electronic Data Processing.

EEPROM or **E^{2}PROM** Electronically Erasable Programmable Read-Only Memory.

Electrical Test See **Wafer Sort.**

Electron Beam Lithography Developed primarily for the requirements of VLSI processing, E-Beam lithography allows for sub-micron pattern generation for producing mask plates or maskless lithography.

Elephant Usually a glass or quartz shielding tube used to transport and protect wafers before and after a deposition or diffusion process.

Elipsometer A machine used in measuring the thickness of oxide layers very accurately.

Epi or **Epitaxial Growth** Vaporized silicon deposited on a wafer.

EPROM Electronically Programmable Read-Only Memory.

ESD Electro Static Discharge (static electricity).

Etch Generally, the chemical removing of a material by bathing in acid. Also see **Dry Etch.**

Evaporation The vaporizing of a material such as aluminum or gold and subsequent depositing of the vapor on the wafers. Also called evap.

Expose In masking after proper alignment of mask to wafer, light is allowed to activate or polymerize the photoresist on the wafer, much like exposing film in a camera.

FAB Fabrication, i.e., wafer fabrication area is called FAB or Wafer FAB.

FET Field Effect Transistor.

FI or **Final Inspect** The last inspection that is given to a crystal, wafer or die to insure that it is within the specifications required.

FIFO First In/First Out.

Final Test The last check of quality for electrical characteristics prior to a package part being shipped to the customer.

Flat Ground along one edge of the crystal prior to slicing, the flat is used as a point of reference during fab processing and as an orientation point for scribing and breaking the die into individual circuits.

FPGA Field Programmable Gate Array.

FPLA Field Programmable Logic Array.

Front End In microelectronics, usually refers to the wafer fabrication process as a whole.

Furnace Generally refers to high temperature cylinders used for depositions and diffusions in wafer fab. Crystal growing machines are also referred to as furnaces.

GaAsP Gallium-Arsenide-Phosphide. A semiconducting material from which LED chips are made.

Gate Array See **Array.**

Germanium A chemical element sometimes used in place of silicon as a semiconductor substrate. Symbol-Ge.

Hard Bake Generally, in masking, the baking of wafers at low temperature to remove moisture and provide for better adhesion of the photoresist after develop and prior to etch.

Hardware Computer systems articles made of metal (i.e., frame, cabinets, etc.) versus the systems SOFTWARE or programming components.

Hermetic Usually refers to hermetically sealed IC packages; they are sealed such that air (or anything else) can neither get in nor out.

HF See **Hydrofluoric Acid.**

Hi Rel High Reliability.

HMDS Hexamethyl disilazane (used at prime in masking).

HMOS High Density MOS

Hybrid Generally, the combining of IC's with one or more discrete devices packaged together.

Hydrofluoric Acid A commonly used acid which is used to etch oxide from silicon wafers.

Hydrogen Colorless, odorless, highly explosive gas used in a variety of fabrication processes.

I²L Integrated-Injection Logic.

IBM Ion Beam Milling (See **Dry Etch**).

ID Inside Diameter.

I/O Input/Output.

Ingot See **Crystal.**

Inked Die Die which do not pass the wafer sort test and are marked with a drop of ink to show they are rejects.

Integrated Circuit The placing of several components (i.e. transistors, resistors, capacitors, etc.) on a single chip. Abbreviated: IC.

Ion An electrically charged atom or group of atoms, the electrical charge of which results when a neutral atom or group of atoms loses or gains one or more electrons.

Ion Impant A method of magnetically segregating specific dopant atoms and then accelerating them so that they "blast" into a wafer. Used as a replacement for furnace depositions where critical control of small amounts of dopant are necessary.

ISL Integrated-Schottky Logic.

JFET Junction Field Effect Transistor.

Jig See **Alignment Jig.**

Junction The area where N and P type materials in a semiconductor come together.

Jungle Generally, the menagerie found at the back end of diffusion or deposition systems, the entire collection of tubes, lines, bubblers, injectors, etc.

Laminar Flow Laminated filtration systems used through virtually all processing operations to control air velocity and prevent the introduction of particle contamination.

Laser Light Amplification by Stimulation of Emission of Radiation.

Lattice Generally refers to the organized atomical structure of wafers or crystals.

LCD Liquid Crystal Display.

Leads The prongs or connecting areas of a package.

LED Light Emitting Diode.

LIFO Last in/First out.

Linear A circuit whose output is an amplification of its input.

Lint Free Clothing or paper, usual contamination sources, are now made to minimize the amount of lint contamination introduced into the fab area.

LOT See **Run.**

LSI Large Scale Integration.

LTO Low Temperature Oxide.

Masking The fab process whereby each layer of the process is photographically transposed onto the wafer so that a deposition can be accurately placed within selected areas of the circuit.

Mask Plate A thin, extremely flat square of glass, quartz or other material, on which the circuit pattern is printed to be used in transferring the design of the circuit to the wafer.

Mask Set The group of mask plates, which when processed in the proper order, help build the circuits of a wafer, one layer on top of another. There can be from three to fifteen mask plates in a mask set.

Melt The liquified poly-silicon charge from which a single structured silicon crystal will be grown. Also refers to the liquified source (i.e., aluminum) in an evaporation system.

Memory Generally refers to circuits which have the ability to store information which can be recalled at a later time.

Metal Gate See **Silicon Gate.**

MIC Microwave Integrated Circuit.

Microcircuit Refers to any small solid state electrical device. Another word for integrated circuit.

Micron Equal to 1 millionth of a meter. Used in measuring thickness of material at various steps of processing.

Mil Equal to .001 of an inch. Used in measuring thickness, width and depth at various steps of processing.

MIS Management Information System(s).

Misalignment In masking refers to the inaccurate positioning of one mask layer to another.

MMOS Memory Metal Oxide Semiconductor.

MNOS Metal Nitride Oxide Semiconductor.

Modem Modulation/Demodulation. A device that is used to link a terminal or a computer to another computer via the telephone system for the purpose of information exchange.

Mono A prefix meaning one or single, i.e., monosilicon means single structured silicon.

MOS Metal Oxide Semiconductor.

MOSFET Metal Oxide Semiconductor Field Effect Transistor.

MPU MicroProcessor Unit.

MSI Medium Scale Integration.

Nanosecond Equal to one billionth of a second, it is used to measure the speed at which the functions of the circuit work.

Nitride Silicon and nitrogen combined to form an insulation layer on a circuit.

Nitrogen Colorless, odorless gas used in a variety of fab processes usually as a carrier gas, or to maintain ultra clean environment for operations.

NMOS N-channel Metal Oxide Semiconductor.

Nonvolatile Memory A memory circuit which will not lose its data when power is lost.

NPN Negative-Positive-Negative.

N-Type A material or dopant having a negative charge, i.e., phosphorus is an N-type dopant.

OD Outside Dimension.

Optoelectronics The technology which mixes solid state electronics and optics.

Oxidation The process which combines oxygen and heat with a silicon wafer in a furnace to produce a layer of silicon dioxide (oxide).

Oxide Silicon dioxide. Grown on a wafer, oxide is used as a deterrent to dopant penetration in deposition and diffusion processes. Also used as part of the structure of the circuit.

Oxygen A chemical element (symbol: O), it is used extensively in wafer fabrication, i.e., oxide layers.

Package The container which secures and protects a finished chip. It comes in many forms, sizes, and materials and will be plugged into the sockets of the customers' product.

Pads Bonding pads of each chip which will be used to attach the chip to the package leads.

PAL™ Programmable Array Logic.*

Passivation Usually a silicon dioxide or silicon nitride layer put over an existing layer of the wafer to protect against moisture and contamination.

Passives Semiconductor devices that have no "gain" (i.e., resistors and capacitors).

Pattern Generator Optical or E-Beam tool used to make the mask plate reticle.

PCB Printed Circuit Board.

PDS Plannar Diffusion Source. Solid disks of doped material that are alternately placed with the silicon wafers on a sled and processed through furnaces; a method of deposition.

PECVD Plasma Enhanced Chemical Vapor Deposition.

Pellicle A protective covering that adheres to a mask plate, it allows the plate to be cleaned without disrupting the integrity of the pattern and thus the defect level will remain low.

Phosphorus A chemical element (symbol: P) used as an N-type dopant in such processes as crystal growing, epitaxy, ion-implant, and numerous furnace depositions.

*****PAL** is a registered trademark of Monolithic Memories, Inc.

Photolithography Masking.

Photoresist A liquid, photoresist is spun on a wafer which will be activated by ultraviolet light at expose in the masking process. Similar to film in an ordinary camera in its sensitivity to light.

Pin Hole Defects caused by contaminates on the wafer, mask or in the photoresist.

PLA Programmable Logic Array.

Planar Process A patented Fairchild Semiconductor process for making various semiconductors and IC's using the monolithic processing technique of building the circuits in a layering manner, one on top of another.

Plasma Etch See **Dry Etch.**

PMOS P-Channel Metal-Oxide-Semiconductor.

PNP Positive-Negative-Positive.

Polishing The process whereby a mirror-like finish is put on sliced wafers.

Poly A prefix meaning *many*, i.e. poly-silicon contains many different crystal orientations. Also refers to the process in which vaporized silicon is deposited on a wafer in such a way that it does not grow in a single (mono) structured form.

Predep Predeposition, actually a misnomer which has evolved meaning deposition. Generally in high temperature furnace applications.

Preform Usually refers to a sliver of pure gold used in attaching a chip to a package on larger sized chips.

Prime A masking procedure which helps promote better adhesion of the photoresist to the wafer.

Probe Meaning varies from area to area. Can mean the same as wafer sort, or the fine wire which contacts the wafer in order to read the electrical characteristics of the circuit.

Probe Card In wafer sort, the card containing several probes which is used to automatically check each circuit on a wafer for acceptable electrical function.

Process Specification The method or "recipe" established by the engineering staff for doing a given operation. Also called a Spec.

Profile Generally refers to the temperature readings measured over the length of a furnace tube in processing.

Projection/Proximity Masking exposure methods in which the wafer and mask plate have no contact, thus lengthening the mask usage due to less contamination of the mask plate.

PROM Programmable Read-Only Memory.

Proprietary A device, process, or equipment that has been developed by a company for its exclusive use (or the company's licensee).

Prototype A model or first thing of its kind.

P-Type A material or dopant having a positive charge, i.e., boron is a P-type dopant.

Pyrox Wafer fab operation for oxide deposition.

QA Quality Assurance.

QAP Quality Assurance Procedure.

QC Quality Control.

RAM Random-Access Memory.

R&D Research and Development. Generally where new products, processes, and equipment are started and developed.

Real Time Generally refers to the continual monitoring of a process or operation so that the changes taking place can be guided or corrected while in process.

Resist See **Photoresist.**

Resistivity The property of electrical resistance in a material. (Inverse of conductivity.)

Reticle A mask plate that has only one of the components that will be printed on a wafer. It may take many reticles to produce an actual "working mask plate" (i.e., one reticle each to produce a plate that has alignment marks, test die, process monitoring devices and the actual active circuit pattern).

RF Radio Frequency.

RIE Reactive Ion Etch. (See **Dry Etch**).

ROM Read-Only Memory.

Route Slip See **Run Card.**

RTL Resistor-Transistor Logic.

Run Card Generally refers to the paper work which accompanies a group of wafers (or crystal section, or finished units, etc.) through processing. Used as a method of controlling product movement and provides historical data for measuring effectiveness of process and design.

Schottky A bipolar product which attains higher speed of operation than TTL.

Scribe The first step in assembly, a wafer is separated into individual chips by a laser cutter, saw, or diamond point.

Second Source Usually more than one company will make the same device. This is for the protection of the customer in case a catastrophic problem at one manufacturer prevents them from being able to supply a certain device.

Seed In crystal growing a piece of single structured silicon which upon contact with the melt (molten poly-silicon) starts a crystal to be grown which has same structure as that of the seed.

SEM Scanning Electron Microscope. Used in examining portions of a circuit by allowing the viewer to see an image as much as 15,000 times its actual size.

Semiconductor A material such as silicon which falls between conductors (such as copper) or nonconductors (such as glass). Generally refers to integrated circuits or other solid state devices formed in semiconducting materials during manufacturing.

Silicon An element (symbol: Si) found in large quantities on earth in sand, used as the major material in which integrated circuits are built because of its ability to regulate the conduction of electrical current.

Silicon Dioxide See **Oxide.**

Silicon Gate Uses silicon for the gate of the transistors in MOS processing (contrasts with Metal Gate, another MOS process).

Silicon Valley Generally refers to the scenic Santa Clara Valley of California because of the numerous semiconductor manufacturing companies located there.

Silicon Wafer See **Wafer.**

Single Structure In silicon crystal it refers to the way in which silicon atoms grow in the exact same manner one on top of another and side by side.

Sled A vessel, usually made of quartz or silicon, used for holding wafers during furnace processing.

Slice Refers to one silicon wafer (or other material) which has been cut from a crystal. Also, to a wafer which has been polished and is "in process" through the manufacturing area.

Slicing The cutting of a silicon crystal in a saw in order to make wafers on which IC's will be made.

Socket The apparatus which a finished, packaged, IC will be plugged into in order for it to help operate a customer's product (TV's, computers, watches, etc.). Also refers to similar apparatuses for testing these parts.

Soft Bake or **Post Spin Bake** After resist is deposited on the wafer it is baked in a low temperature oven to remove any solvents and give better adhesion of the resist prior to alignment.

Software See **Hardware.**

SOI Silicon On Insulator.

Solid State Generally used to differentiate between the old, bulky, sluggish, and slow vacuum tubes and the new, small, efficient and fast integrated circuit.

Sort or **Wafer Sort** The testing of each die on a wafer upon completion of fab processing to determine and mark the acceptable die from the non-acceptable die (the non-acceptable die are marked and will be discarded).

SOS Silicon-On-Sapphire.

SPEC See **Process Specification**, also refers to customers' written needs in terms of size, package configuration, and electrical characteristics.

Spin The operation in masking whereby photoresist is applied to a wafer and rotated at high speed so that the photoresist coats the wafer with a uniform film.

Sputter Method of depositing various types of thin films on wafers by ion bombardment of a target.

SSI Small Scale Integration.

Starting Material See **Substrate**.

Stepper or **Step-and-Repeat** In making mask plates a step-and-repeat camera is used to transform the pattern image of the reticle onto the surface of the plate. In some fab processing, a stepper is used to project the reticle's image directly onto the resist spun wafer and does not employ a mask plate (also called DSW for Direct-Step-On-The-Wafer).

Strip In fab, refers to the stripping of the photoresist after etch usually in a wet chemical bath or in a plasma chamber.

Substrate Refers to the wafer on which a product will be built. Also called a wafer or starting material.

Target Used in sputtering, a target may be composed of various materials (aluminum, aluminum alloys, gold, etc.). Under vacuum these materials are bombarded by ions which allows the materials to then be deposited on the wafers within the vacuum chamber.

Test Generally refers to the final electrical and cosmetic evaluation of a product before it is shipped to a customer.

Test Die Usually designed into a mask set, a test die will occupy a small percentage of a wafer and can be used for easier and more complete means of evaluating an entire wafer's quality without damage to the actual product dice.

Thermocouple Used in various processing equipment for the purpose of obtaining very accurate temperature measurement. (See **Profile**).

Thin Film In wafer processing the depositing of any number of materials on the wafers via sputtering, ion-implant, vacuum evaporation, or E-Beam. Typical materials that might be deposited include metals, poly (poly-silicon), nitride, silicon-chromium, etc. Usually these materials will only be a few hundred angstroms in thickness.

Torr In vacuum systems the remaining pressure inside the chamber after pumpdown is a measure of atmospheric pressure expressed as Torr. (Torr = 1/760 of atmospheric pressure).

Toxic A poison. Many of the chemicals used in manufacturing can be dangerous if not used in the proper and specified manner.

Transistor A solid state device used to amplify or switch electrical current.

TTL Transistor-Transistor Logic.

Tube Most generally refers to quartz or silicon cylinders in which high temperature depositions or diffusions take place.

ULA Uncommitted Logic Array.

Undercut In masking, generally refers to the material directly under the photoresist that is inadvertantly etched away.

Units Generally refers to a group of packaged, finished IC's.

UV Ultraviolet.

Vacuum Practically, a chamber with very low pressure ideal for depositing various dopants and metals on silicon wafers with great uniformity.

Vapox Widely used to mean the protective coating of silicon dioxide (also called "glass") which covers the circuits of a wafer in the final stage of wafer processing.

VHSIC Very High Speed Integrated Circuit.

Viscosity Generally refers to the thickness of a fluid, i.e., blood is more viscous than water.

VLF Vertical Laminar Flow.

VLSI Very Large Scale Integration.

VLSIC Very Large Scale Integrated Circuit.

VMOS Metal Oxide Semiconductor that is made with a "V shaped notch" instead of in a flat horizontal plane.

Wafer The silicon disc slice from a crystal on which integrated circuits are manufactured. Also called a substrate or starting material.

Wafer Fab The area in which circuits are manufactured, it usually consists of masking, diffusion, deposition, and other operations which will transform a polished wafer into hundreds of chips.

Wafer Sort See **Sort.**

Window The opening of minute holes in the oxide (or other surface material) of a wafer during the masking process. Generally for the purpose of depositing other materials (dopants, oxide, metal, etc.) at the next processing step.

Working Plate See **Mask Plate.**

X-Ray Lithography Uses x-ray rather than optical or E-Beam for generating mask plates or for image transfer directly to the wafer.

Yield The amount of good things as compared to the total possible good things, i.e., on a wafer which has 100 possible chips and 65 are found to be good then the yield = 65%. Or if a "run" of wafers has 50 wafers to start and 41 wafers are finished the run has a yield of 82%.

Common Symbols and abbreviations used:

$>$	Greater than
$<$	Less than
$\sim$	Approximately
Å	Angstrom
$F°$	Fahrenheit Degrees
$C°$	Centegrade Degrees
μ	Micron
Ω	Ohm
ρ	Resistivity
β	Beta (gain)
Cm	Centimeter
$\Omega/\square$	Ohms per square
t	Thickness
$\pm$	Plus or minus

X-TAL - Crystal

NOTES

NOTES: